To Care for Creation

To Care for Creation

The Emergence of the Religious
Environmental Movement

STEPHEN ELLINGSON

The University of Chicago Press
Chicago and London

Stephen Ellingson is associate professor of sociology at Hamilton College.

The University of Chicago Press, Chicago 60637
The University of Chicago Press, Ltd., London
© 2016 by The University of Chicago
All rights reserved. Published 2016.
Printed in the United States of America

25 24 23 22 21 20 19 18 17 16   1 2 3 4 5

ISBN-13: 978-0-226-36724-8 (cloth)
ISBN-13: 978-0-226-36738-5 (paper)
ISBN-13: 978-0-226-36741-5 (e-book)
DOI: 10.7208/chicago/9780226367415.001.0001

Library of Congress Cataloging-in-Publication Data

Names: Ellingson, Stephen, 1962– author.
Title: To care for creation : the emergence of the religious environmental
    movement / Stephen Ellingson.
Description: Chicago ; London : The University of Chicago Press, 2016. |
    Includes bibliographical references and index.
Identifiers: LCCN 2015045697 | ISBN 9780226367248 (cloth : alk. paper) |
    ISBN 9780226367385 (pbk. : alk. paper) | ISBN 9780226367415 (e-book)
Subjects: LCSH: Environmentalism—Religious aspects. | Ecotheology. |
    Nature—Religious aspects. | Human ecology—Religious aspects.
Classification: LCC BT695.5 .E576 2016 | DDC 201/.77—dc23 LC record
    available at http://lccn.loc.gov/2015045697

⊚ This paper meets the requirements of ANSI/NISO Z39.48-1992
(Permanence of Paper).

To Zach

CONTENTS

ACKNOWLEDGMENTS

This book took much longer to write than I anticipated, and I accrued numerous debts to those who helped me along the way. First I must thank the interviewees who graciously gave their time and thoughtful answers to my questions. Many sent me additional documents and pointed me toward other REMOs to include in the study. A small number of interviewees read a rough draft of the manuscript in 2014 and offered helpful comments. The Louisville Institute funded a year of data collection during the 2007–2008 academic year (grant #2007004). Hamilton College provided two sabbaticals at the start and end of the project without which the book would not have been possible. The Arthur C. Levitt Center at Hamilton College funded a summer research project in 2011 that jump-started the analyses for chapters 2 and 4 and gave me the opportunity to present my findings to colleagues across the college on two occasions.

I benefited from the help of numerous individuals as well. Chris Ansell, Chris Henke, Richard Seager, and Rich Wood provided methodological advice or suggestions about religion and the environment at the outset of the project. I would not have been able to write chapter 5 without the network analysis conducted by Vernon Woodley and Tony Paik that was the basis for our coauthored article in the *Journal for the Scientific Study of Religion* (June 2012). Marcia Wilkinson efficiently transcribed all of the interviews. Several extremely capable undergraduate students coded data and wrote annotated bibliographies (Alexandra London, Sarah Boole, Katie Axelrod, Dan Rudel, Emma Leeds) or conducted preliminary analyses about the emergence of the movement and the development of a green religious ethic (Will Rusche and Andrea Wrobel). My colleagues in the Sociology Department—Yvonne Zylan, Jaime Kucinskas, Jenny Irons, Dennis Gilbert, Ben Dicicco-Bloom, and Dan Chambliss—read the manuscript and provided telling critiques, helpful suggestions, and encouragement. Phil Devenish carefully read the manuscript and pushed me to clarify my arguments in numerous ways. Stacey Himmelberger reformatted all of my notes, figures, and bibliography, and copyedited the manuscript. Bret Olson reworked the two figures from chapter 5 in Photoshop. Robin Vanderwall, my department's administrative assistant, provided invaluable support in countless ways throughout the course of the study. All of these members of the Hamilton College community deserve my thanks. At the Press, Doug Mitchell provided much encouragement, support, and sage counsel about the revisions to the book. His assistant, Kyle Wagner, patiently and efficiently responded to my numerous queries, and the reviewers of the manuscript offered extremely helpful suggestions to improve the book.

Finally, I wish to thank my family for tolerating my many lapses of attention during the months of writing over the past two years. My partner, Jennifer DeWeerth, good-naturedly read the manuscript twice and caught many typos, grammatical mistakes, and poorly constructed sentences. My youngest son, Mesafint, pulled me away from the project and onto the soccer field (or at least to soccer practices) several times a week for the past few years, which has been a welcome diversion. The intrusion of cancer into the life of our family nearly derailed the book altogether. On April 19, 2010, my oldest son, Zach, was diagnosed with leukemia, and I stopped working on the project. For the next three years Zach endured daily chemotherapy, numerous blood transfusions and spinal injections, and the awful physical, emotional, and psychological effects of cancer treatment. Throughout the ordeal he taught me what courage is, how to endure more patiently, how to love more fiercely, and how to be a better human being. He has brought more smiles and tears into my life, and he has inspired me in more ways than he knows. Today, Zach is healthy and in remission and I dedicate this book to him.

ONE

A Greener Faith

From 2007 to 2008, I interviewed more than sixty religious environmentalists who were establishing a new social movement. Near the end of the interview period, I had the opportunity to spend the day with Paul Gorman, the founder and executive director of the National Religious Partnership for the Environment (NRPE). Throughout the early 1990s, Gorman organized a series of conversations between religious leaders from different communities, and in 1993, the NRPE was established. The NRPE is a coalition of national Jewish, Evangelical, Catholic, and mainline Protestant religious organizations dedicated to environmental activism. Gorman had a front-row seat—really a driver's seat—during the formation of the religious environmental movement. He was reluctant to use the word "movement" and instead offered a number of alternative terms—"awakening," "paradigm shift," "renewal," "reconciliation"—because movement did not seem to capture adequately the radical changes to religion and environmentalism that the NRPE hoped to initiate. Gorman explained that during the early conversations about establishing the NRPE, they decided that their goal would go beyond replicating the old forms and practices of religious activism. "We weren't trying to create another interfaith organization or a new organization about human rights or peace. . . . Some of us were very aware of an obligation to set the foundation for this and not be another trendy issue or movement or something that people grab onto as a flavor of the month or a year or a decade." He went on to note that they weren't simply trying to create a new reform movement like the Social Gospel movement of the late nineteenth and early twentieth centuries as a way for liberal and justice-oriented religious groups to stay relevant. He claimed that their endeavor was something new altogether. Then he sat up on the couch in his living room, leaned forward and in a voice filled with emotion said:

We're not the environmental movement in prayer. We're trying to bring care for God's creation to the heart of religious life, or to weave programs to care for God's creation across the entire path of religious life. And henceforth, to be religious means if you love your neighbor, you care for creation. . . . Care for creation brings life to faith, it revivifies faith. Wasn't this the idea in the first place? Isn't God's creation and God's revelation calling us into the most intimate relationship? . . . To be one with God, we must be part of God's creation. That's the deal; that's the original deal.

At this point of the interview, I had a something of an epiphany—Gorman's comments alluded to God's original designs for humankind in the Garden of Eden and the hope of restoring the divine-human relationship that lies at the heart of the Judeo-Christian traditions. This new movement was as much about the renewal of faith as it was about saving the environment. On my interview guide I had jotted down my insight, "not the environmental movement in prayer; not the shock troops for an embattled and rudderless green movement at large." Religions were not simply trying to persuade Congress to enact environmentally sound legislation, nor were they eager to join hands with the secular movement to slow global warming. Instead, they were hoping to forge a new wave of environmentalism and in doing so, to renew and deepen religious faith.[1] In one of the documents Gorman forwarded to me after our interview, "God's Climate Embraces Us All," the NRPE stakes out its mission more formally, clearly distinguishes religious environmentalism from the secular movement, and summarizes the organization's goal of revitalizing faith by integrating it with environmental activism:

How can religious life help make climate change a moral issue? Deep religious values are having fresh power for people of faith in that first generation to behold the whole earth, precious and in peril. In Genesis, God designates creation as "very good" (Gen. 1:31) and commands us to "till and to tend the garden" (Gen. 2:15). In Psalms we read, "The Earth is the Lord's and the fullness thereof" (Ps. 24:1). We have a paramount obligation to care first for the "least of these" (Matthew 25:35) and to assure the future well-being of all life on Earth in God's "covenant which I make between you and every living creature for *perpetual generations*" [italics original]. These values are flowing vividly into worship, sermons, and religious observance. They're inspirations, not frames, living articles of faith, not talking points. They are the reasons why care for God's creation is becoming the most compelling new cause in religious life and why global climate change is the issue that is most moving people of faith to act.

When I returned home, I reread the interview transcripts and quickly discovered that most of my informants had also spoken about the movement in similar terms. Their goal, like that of the NRPE, is to awaken religious congregations and/or individuals to environmental problems and to take action because it is a sacred duty. Religious environmental movement organizations (REMOs) put forth the claim that environmentalism is integral and necessary for an authentic and meaningful life of faith. The point of religious environmentalism is to deepen and enrich the lives of the faithful and then to ameliorate ecological problems. I began to see religious environmentalism in a new way. It is not another social movement, nor is it simply an effort to renew or revitalize religion, which is how scholars have historically treated religious social movements.[2] It is a movement that attempts to renew faith and to mobilize the faithful to save creation. But just what is novel about religious environmentalism in comparison to older religious social movements? Are REMOs fundamentally different from secular environmental movement organizations, or are they simply a variation of the postindustrial, new social movements (NSMs) that stress identity and nonmaterialistic goals?[3]

In many ways the new religious movement does not follow the model of the Sierra Club or other major national environmental organizations. REMOs that participated in the study are not staffed by professional activists that rely on lobbying, litigation, or legislation to save nature. They are clergy and laypersons, deeply enmeshed in specific religious communities that are engaged in a novel attempt to mobilize and change the culture of American religions by placing concern for the environment at the heart of faith. They do not use the language of science or politics to describe their work, but instead they use the idioms of stewardship and justice, covenant and redemption that are central to many religious traditions in the United States. They do not offer gloom-and-doom scenarios of ecological destruction or technical discussions of how to reduce greenhouse gas emissions. Instead, they offer a hopeful vision of how God wishes to save all of the creation, and they articulate a call to respond to God's commandments and promises about nature.[4] REMOs focus on awakening religious groups and individuals first before solving specific ecological problems. They tend to work on ecological issues that have a clear impact on human beings, and thus they downplay traditional environmental goals of protecting wilderness areas and endangered species. They eschew long-term ties to secular environmental organizations and are even wary of entering into short-term projects with their secular counterparts. Instead, they have developed a parallel movement, replete with a unique set of religious goals and strategies, culture, and collective identities. Finally, and perhaps most importantly, REMOs have shown members of America's religions that environmentalism is congruent with and even mandated by religion itself. They have begun persuading individuals and groups that to be fully and authentically religious also means that one must be an environmentalist.

Why have religions taken up the cause of environmentalism? The environment has not been an active concern among most religious organizations in the United States, and secular movement organizations have made few efforts to ally with national religious bodies or local congregations. However, during the early 1990s, activists from a variety of religious traditions established a small number of independent, nonprofit religious environmental organizations, and since 1997, more than seventy such groups have been founded. Collectively, these organizations have launched a new religious environmental movement and have gained an institutional foothold within American religion. REMOs have created a new green religious ethic and culture where none had existed and launched programs to make religious buildings environmentally friendly, fight toxic waste and mountaintop removal, protect watersheds, and promote sustainable agriculture.

The successful emergence of this new movement is no small accomplishment given the number of institutional challenges the organizations face. Religious environmentalists are trying to mobilize an indifferent, sometimes hostile, and certainly distracted audience. The environment is not a core issue within any American religion, and as multi-issue organizations, denominations and congregations are fully engaged in a variety of human-centered social activities (for example, food pantries, relief, and development work). In addition, activists have few allies within denominations or other religious bodies. Conservative Protestants and Catholics tend to equate environmentalism with antireligious liberalism and have expressed little interest in supporting new initiatives on the environment, particularly if those initiatives will take them away from their core interests in evangelism and the public issues surrounding the cultural wars. During the 1990s and early 2000s, internal fights over sexuality, membership declines, and persistent revenue shortages consumed liberal and mainline Protestant groups, and thus there was little energy and even fewer resources within these groups to support a new environmental initiative. Moreover, poll data suggest that Christians' attitudes about the environment and green practices have not strengthened since the 1990s, and that non-Christian and nonreligious Americans continue to hold stronger proenvironmental beliefs and attitudes than Christians.[5] Other studies suggest that liberal Protestants and Catholics tend to hold stronger proenvironmental beliefs than conservatives, but there is no compelling evidence to suggest that any particular tradition or denomination within American Christianity is becoming "green."[6]

Until recently, American religions had given little attention to developing a theology or ethics about the environment, and the small amount of work that had been done by scholars and denominational officials about faith and ecology was largely unknown at the congregational level or ignored within particular national churches. In order to gain a hearing and then garner the legitimacy and support necessary to become established, activists must create new green religious traditions and a new collective green religious identity. They must persuade people who have historically believed they are ethically required to care for human beings that it is also religiously necessary and legitimate to care for nature. In short, REMOs faced strong barriers to legitimacy, they found themselves with few resources on which to draw, and they faced audiences that would be difficult to mobilize.

The rise of religious environmentalism raises intriguing questions about movement emergence, the development of a new type of movement organization, and the creation of new religious traditions. This book explains how religious environmentalism emerged despite various institutional and cultural barriers, and why the new movement organizations follow a logic and set of practices that set them apart from the secular movement. Drawing on interviews with leaders of sixty-three REMOs, the book tells three new stories. First, it is an account of how entrepreneurial activists tap into and improvise on a variety of theological, ethical, and symbolic traditions in order to create a compelling call to arms and strategies to mobilize religious audiences. It also shows how activists work within the limits imposed on them by their own religious communities and the paucity of resources to create a new type of movement organization and launch the new movement. While there are a few scholarly monographs on the US religious environmental movement and a handful of articles, they do not provide a very compelling and complete answer to the questions of how and why the new movement emerged and developed. Therefore, this book offers a deep and comprehensive description and explanation of this potentially important new wave of environmentalism in the United States.

Second, the book develops a cultural-institutionalist perspective of movement emergence and institutionalization that incorporates the concept of "embeddedness." The most common sociological explanations for movement emergence do not fit the case of religious environmentalism very closely. The resources mobilization perspective claims that new movements cannot emerge when there are not an accessible and sizeable amount of human, financial, technical, and moral resources. Yet REMOs were founded and continue to operate with extremely limited resources. Work within the political opportunity structure (POS) and the more recent strategic action fields approach of Fligstein and McAdam argue that new movements and movement organizations emerge when the political system grows unstable or institutions face some type of crisis.[7] However, REMOs are not oriented toward or interested in working within the formal political system. They act with little regard for the state or changes in leadership within state and federal governments. Moreover, the movement emerged over a roughly twenty-year time period that saw numerous shifts in the control of Congress, judicial rulings on the environment, and a welter of ecological crises, yet its emergence cannot be easily pegged to clearly identified political events or shifts in opportunities. In general, scholars have tended to emphasize the structural factors that are associated with the emergence of new social movements or describe the sociopolitical contexts, events, and crises that compel new groups to engage in protest. However, such approaches do not adequately explain the concrete dynamics and process of emergence or theorize how activists exercise agency within and despite the limits of structure. Thus the case of religious environmentalism poses a theoretical puzzle—how did this new movement emerge with few resources, in a context that was not particularly opportune, and with no clear institutional or political crisis that called for a religious response?

In order to understand how this new movement emerged, I argue that we need to explain how activists and the REMOs they operate address a series of challenges about organizational identity and legitimacy, authority and resources, religious authenticity, and integrity. In particular, I show how such decisions and actions are constrained and facilitated by their embeddedness or deep connections to specific religious organizations, audiences, and systems of meaning. For example, many Christian and ecumenical activists feel pushed to use the language of "stewardship" and to emphasize projects that will have a direct impact on human beings because of the constraints they experience from the communities they were trying to mobilize. These communities lack a historic commitment to the environment and a practical theology or ethic supportive of environmental action. Instead they stress the religious mandate to care for their neighbor. Similarly, evangelicals' deep suspicion about working with non-Christian organizations forces evangelical REMOs to eschew alliances with REMOs that would not be considered Christian by members of the broad evangelical community. In sum, I show how institutional, relational, and cultural embeddedness profoundly shape activists' decisions about every step of the movement's emergence and development—decisions about mission and goals, organizational form, protest tactics, alliance formation, and framing.

Third, this book integrates theories and concepts about cultural innovation and social movement emergence and development. Religious environmentalists face a difficult organizing problem: how to persuade religious individuals and groups that holding a proenvironmental identity and engaging in environmental action are integral and not inimical to the life of faith. I show how activists borrow, adapt, and rework resources from various religious traditions to create new meanings about religion and nature, and the religious person's duty to the natural or created world. Existing scholarship on innovation has identified some of the creative strategies and staked out a contextual argument to explain when innovations are likely to occur, and I build on this body of work by identifying the cultural and structural constraints on the innovative process. More specifically, I show how environmental entrepreneurs are constrained by the expectations (real and assumed) of their key constituencies and by the relationships they have with specific religious organizations. In other words, innovation is limited by the nature of activists' embeddedness within particular religious settings and traditions. Thus the book fine tunes and integrates theories about cultural innovation with social movement emergence and development by demonstrating how creativity and agency are constrained, channeled, and enabled by different types of institutional embeddedness.

The Study

In 2005, I wanted to start on a new project on American environmentalism and began reading histories of the movement. After making my way through several monographs and a significant number of articles, I was struck by the absence of religion in the movement. It seemed odd that religions were not involved in environmentalism given that religions, especially Protestant denominations and congregations, have played central roles in nearly every major American social movement. Local churches, denominations, and parachurch organizations have provided the ideological and material resources, and developed the repertoires of contention, that fueled the antislavery and temperance movements in the nineteenth century, and the peace, civil rights, and pro-life movements in the twentieth century. Religious organizations and their members also have been deeply involved in national movements regarding immigration, nuclear weapons, capital punishment, homosexuality, and, more recently, in a growing number of efforts at the local level to improve neighborhood safety, public schools, and opportunities for affordable housing and job training.[8] Recent surveys of American congregations reveal the central role congregations play in providing social services for the needy and, more generally, in trying to change the world through social services, advocacy, and moral suasion.[9] At the same time, I began reading journalistic accounts of new religious organizations that were petitioning Congress not to drill in the Arctic, becoming involved in public discussions about global warming, or investing in green energy. It appeared that religions were going green, but little scholarly attention had been directed to this new phenomenon.

Intrigued, I conducted a pilot study during the summer of 2005. I interviewed six religious leaders who represented three mainline denominations, two ecumenical parachurch organizations, and one evangelical environmental group. I also spoke with five activists from secular environmental groups whose focuses ranged from wilderness preservation to toxic waste issues. I collected documents—annual reports, newsletter and magazine articles, denominational and parachurch groups' official statements or theological writings on the environment. By the end of the summer, I was convinced that a new religious social movement was developing and decided to make it the focus of my research in the coming years. I compiled a list of politically active or policy-oriented organizations employing faith-based approaches to environmental issues by combing through information from the National Religious Partnership for the Environment, the National Council of Churches Eco-Justice Program, the Web of Creation, the Forum on Religion and Ecology, and the Sierra Club's annotated directory. I also asked each interviewee to name other faith-based environmental organizations to make sure I had not overlooked any REMOs. I excluded organizations with no political or policy agenda because they did not meet the criterion of being a movement organization.[10] Because I relied primarily on published lists to identify REMOs, it is likely that smaller, lesser-known REMOs were omitted. In all, I identified a population of eighty-three REMOs as eligible to participate in the study, and representatives from sixty-three, or 76 percent, agreed to be interviewed (see the appendix).

Between August 2007 and July 2008, I conducted one-hour phone interviews with the executive directors or equivalent representatives of REMOs in the United States.[11] The telephone interviews were recorded and later transcribed. The interview schedule was organized into four broad topics:

(1) REMOs' history, characteristics, and environmental goals and activities; (2) interorganizational ties with other REMOs; (3) ties with secular movement and other religious organizations; and (4) organizational structure, finances, and influence. I supplemented this data with organizational histories, mission and goal statements, and accounts of activities from each REMO's web page and documents my interviewees sent me about their REMO's histories and activities.

A Brief History of Religious Environmentalism

In *The Greening of Protestant Thought*, Robert Booth Fowler notes that "there is no neat chronology, nor is there any simple way to illuminate all the steps towards an engagement with the environmental cause that took place between 1970 and 1990."[12] It is an uneven history that is full of false starts, roadblocks, and short-lived institutional initiatives. Early attempts by academicians during the 1970s and 1980s to create eco-theologies and formal statements drafted by denominational officials were rare and gained little traction. Methodists, Lutherans, and Presbyterians drafted official statements in the 1970s but did not develop programs to put their green ideals into practices. Liberal and mainline thinkers, such as Paul Santmire, John Cobb, and Rosemary Radford Ruether, evangelical biologist Calvin Dewitt, and Catholic theologian Thomas Berry, laid out the biblical and theological rationale for Christian ecology, but their work did not generate enough interest or support to push religions to get involved in any sustained way in the environmental movement.

The early green religious thinkers set out to create a comprehensive moral code or ethic that identifies the value God places on nature and the sacred responsibilities humans have to safeguard the environment.[13] They developed three broad ethics of creation care that would be used and expanded by contemporary activists: stewardship, eco-justice, and creation spirituality. Stewardship is the most common ethic used by Catholic, mainline and Evangelical Protestant, and even some Jewish REMOs, although as Kearns points out, conservative and moderate Protestants are more likely to rely on this ethic than more liberal groups.[14] This ethic emphasizes the biblical mandate found in the second chapter of Genesis to take care of the earth. One common idea that cuts across specific incarnations of the stewardship ethic is "an image of God as the sacred Creator and of all nature—all the material world—as God's creation. This relationship makes nature special and all life sacred and establishes that no one should dare destroy what God has created."[15] In this framework, environmental problems are understood to be rooted in sin or alienation from God. They are the manifestation of disobedience, arrogance, and greed. The solution to environmental problems requires repentance and a commitment to accept the responsibility of preserving the integrity of the natural world.[16] The stewardship ethic tends to locate the environmental problems and their solutions at the individual level and thus does not push religions to consider the systemic or structural causes of environmental degradation or identify how power relations are implicated in environmental degradation. A second ethic, eco-justice, corrects the overly individualistic emphasis of the stewardship ethic. It draws on biblical mandates to care for the poor, the weak, the powerless, and the most vulnerable in society and extends it to the environment. Religious groups are called on to care for and defend those who are most adversely affected by pollution, climate change, or the loss of arable land, for example. Moreover, the ethic focuses activists' attention on understanding and acting to change the systems of power and institutional forces (for example, corporate greed or flawed public policies) that produce a world marred by ecological and social inequality.[17] Creation spirituality, the third ethic, was developed by Catholic thinkers, and it "attempts to reorient people to understand the proper place of humanity as part of a panentheistic creation as opposed to seeing humans as separate from creation and God outside of creation."[18] In other words, creation spirituality rejects religious anthropocentrism and argues that humans and nature are mutual partners in God's cosmic plan. This ethic calls on humans to reimagine their place and role in creation, to see the myriad ways in which they are deeply interconnected with nature, and then to act in ways to preserve and enhance the webs of interdependence that bind the human and nonhuman worlds.

These three ethics were developed and put into practice during the late 1980s and early 1990s as national denominations and religious leaders began issuing statements about the environment and sending delegates to a series of international meetings about religion and the environment.[19] During the first five years of the 1990s, there was a flurry of activity around religion and the environment. In 1990, Pope John Paul II delivered a World Day of Peace message titled "Peace with God the Creator, Peace with All of Creation" in which he acknowledged the reality and dire consequences of climate change, called the ecological crisis a "moral obligation," and urged all people "to recognize their obligation to contribute to the restoration of a healthy environment."[20] This was followed in 1991 by the United States Conference of Catholic Bishops' pastoral letter "Renewing the Earth."

The Presbyterian Church, USA issued a statement on restoring creation in 1990, and similar statements were crafted by the Episcopal Church in 1991, the American Baptist Church in 1989 and 1992, the Evangelical Lutheran Church in America in 1993, and a group of 150 evangelical leaders, who signed "An Evangelical Declaration on the Care of Creation" in 1994. In terms of the nascent movement, a key catalyst was the 1990 "Open Letter to the Religious Community" (signed by thirty-four internationally recognized scientists) that called on religions to bring their moral authority to bear on environmental issues in order to save the earth. Carl Sagan, Al Gore, and Paul Gorman of the National Religious Partnership for the Environment (NRPE) organized a series of tradition-specific consultations during 1991 and 1992 to help religious leaders develop a theological response to the "Open Letter."[21] In my interview with Gorman, he recalled how these various consultations were a vital first step in mobilizing religion: "we had to get the theology right because we weren't even going to get in the door, past gatekeepers or skeptics, if we didn't do that." On the heels of these early meetings, Gorman helped create the NRPE in 1993, and a handful of other REMOs organized in the first half of the decade. The pace of REMO formation picked up after 1996, and the NRPE proved to be a "movement midwife" as it funded a number of specific programs through its member organizations: the United States Conference of Catholic Bishops (USCCB), the National Council of Churches (NCC), the Coalition on the Environment and Jewish Life (COEJL), and the Evangelical Environmental Network (EEN).[22]

Despite an early focus on national issues, the REMO field also expanded at local, state, and regional levels.[23] Many REMOs were founded in response to place-specific environmental issues, such as supporting energy deregulation in California (Episcopal Power and Light, 1996, which became the Regeneration Project) or preserving the Chesapeake Bay watershed (Chesapeake Covenant Congregations, 2006). Others were established by entrepreneurial religious leaders who gathered local, state, or regional leaders (Faith in Place, 1999; Religious Witness for the Earth, 2001; Baltimore Jewish Environmental Network, 2006) or by members of particular denominations who wanted to raise awareness about environmental interests within a denomination (Quaker Earthcare Witness, 1987; UU Ministry for Earth, 1989; Presbyterians for Restoring Creation, 1995). The movement gained momentum within evangelical and mainline Protestant churches through a number of high-profile events, such as the 2002 "What Would Jesus Drive Campaign" of the Evangelical Environmental Network and the Regeneration Project's 2005 campaign to screen *An Inconvenient Truth* for thousands of congregations across the United States. In 1990, there were nine REMOs in the United States, but by 2010, there were more than eighty, and they operated in nearly every state.

A Snapshot of the Religious Environmental Movement

Table 1.1 provides a statistical overview of the sixty-three organizations that participated in the study. As noted above, the majority of REMOs were founded after 1996. In terms of religious affiliation, most are either interfaith (41 percent) or ecumenical organizations (27 percent). Mainline Protestant and Jewish organizations each composed 8 percent of the sample, Evangelicals made up 6 percent, and the remaining groups—Catholics, Buddhist, and Eco-Spirituality organizations—each were 3 percent. The most popular environmental issue was climate change (89 percent), which is unsurprising given that data collection occurred shortly after the release of *An Inconvenient Truth*. Other prevalent environmental issues included alternative lifestyles (52 percent),[24] land and water stewardship (40 percent), water pollution control (40 percent), energy conservation (38 percent), and food sustainability (32 percent). In contrast to many secular organizations, REMOs rarely work to protect wildlife, which appears to follow from the strong human-centered focus of most religions' general social ethics. Similarly, few REMOs rely on the tactics of lobbying, litigation, or legislation that are the mainstays of the secular movement. Two are involved in litigation, and while thirty-nine (or 62 percent) engage in advocacy, their 501c3 status severely curtails this work. Most REMOs participate in the "annual religious lobby day" with their state legislatures and occasionally encourage their members to write letters to Congress or state lawmakers. Only twelve (or about 20 percent) engage in direct protest activities. Most REMOs have limited resources as over half have operating budgets of less than $100,000.

Overall, this is a young and resource-poor movement. Its members do not share a common set of frames, goals, or protest strategies. Coordinated and joint action among REMOs is episodic and even rarer between religious and established secular movement organizations. Most REMOs operate at a local or regional level with limited knowledge of the rest of the movement and struggle to coalesce into a more unified and powerful social force. At the same time, leaders of the organizations in the study see themselves as part of a larger process of religious awakening and social renewal. They all believe climate change is the dominant ecological threat to the human and natural worlds, and they aim to mobilize their religious communities to take action. But they recognize that, at this nascent stage, most of their energies must

1.1 Descriptive statistics of REMOs (N = 63)

Variables	%
Year of founding	
1970–1996	43
1997–2007	57
Religious affiliation	
Interfaith	41
Ecumenical	27
Mainline Protestant	8
Jewish	8
Evangelical	6
Catholic	3
Buddhist	3
Eco-Spirituality	3
Theological frames	
Eco-Justice	83
Stewardship	73
Eco-Spirituality	27
Environmental issues	
Climate change	89
Simple or alternative lifestyle	52
Land and water stewardship	40
Water pollution	40
Energy conservation	38
Sustainable food production	32
Air pollution	19
Recycling	13
Overpopulation	3
REMOs' individual ties to other organizations	
Ties to nonenvironmental religious group	40
Ties to secular environmental group	44
Organization size: annual budget	
< $100,000	54
$100,000–$500,000	29
> $500,000	17
Scope of operations	
Local, State, and Regional	71
National	29

be focused on articulating a new green religious ethic, winning support for their message, and marshaling resources to mobilize the faithful effectively. Thus the task of this book is to explain how a new movement emerges by drawing on and adapting preexisting religious culture and activating religious networks.

Explaining the Rise and Growth of the
Religious Environmental Movement

The religious environmental movement became established nearly a quarter century after the first Earth Day in 1970 and roughly three decades after
passage of some of the most important environmental legislation of the
twentieth century (that is, the Clean Water, Clean Air, and Wilderness Acts,
1964–67). Why were religions so slow to take up the cause of environmentalism? There is no comprehensive explanation for this question in the literature, but several lines of thought point toward a cultural and institutional
answer. From the start of the contemporary environmental movement in the
late 1960s, secular environmental movement organizations (SEMOs) did
not seek out religious partners and believed religions were implicated in the
perpetuation of environmental degradation.[25] In particular, secular activists
believed Christianity promoted attitudes toward nature that gave humans
ultimate power to control and use nature however they saw fit.

Religions were not interested in working with environmentalists, in part
because they believed secular movement organizations held views about
such issues as population control, gender equality, and personhood that
were directly at odds with religious beliefs. Some religionists believed that
the environmental movement promoted nature worship and biocentrism,
both of which violate biblical teachings about the human-divine relationship and the privileging of humans over nature.[26] At the same time, the
period between 1970 and 2005 found many religions in the United States
focused on a host of other issues that left them with few resources or little
interest in pursuing environmentalism. These included perennial membership loss, schisms and mergers, intrareligious conflicts over sexuality and
gender equality, and, in some circles, a preference to engage publicly only on
"culture war" issues (for example, abortion, pornography, prayer in public
school).[27] This history of mistrust and nonengagement made it difficult
for religions and the secular movement to develop working relationships,
and their strong embeddedness in different institutional fields led them
to develop different goals, identities, and practices related to action in the
public sphere.

Working from a political opportunity perspective, Kearns argues that the
antienvironmentalism of the Reagan administration (1980–88) motivated
religious organizations to take up the cause.[28] Although the argument is
plausible, she does not provide much direct empirical evidence to support
the claim that the political threats Reagan posed to the environment roused
the nation's churches. Nor does she account for why the majority of REMOs

were founded a decade or more after Reagan left office. Curiously, none of the sixty-three organizations that participated in the study mentions Reagan's antienvironmentalism as a catalyst for mobilization, and only three mention former president George W. Bush's antienvironmental policies. In short, a political context argument does not effectively explain the lag between the opportunity or threat arising from the national political scene and the emergence of the movement. (I will address the applicability of the political opportunity explanations more fully in chapter 2.)

Kearns and others also suggest that the religious movement emerged because the secular movement had lost its moral voice. Smith and Pulver note that "religious-environmental groups stepped in to 'fill a void' created by public policies and secular environmentalists who embraced scientific and technical solutions while ignoring values in the 1980s and 1990s."[29] Again, this is a plausible explanation, and REMOs certainly have infused the larger movement with a clear discourse about morality and ethics, but it does not address the significant lag between the rise of religious environmentalism and the time period (the 1980s) when the secular movement deemphasized values and ethics and adopted technocratic and policy-oriented strategies (also known as the "Three L's—lobbying, litigation, and legislation). Thus moral suasion had been muted for more than a decade before religions responded to this alleged void.[30]

In this book, I develop an institutional and cultural explanation for the emergence and development of the new religious environmental movement.[31] Neoinstitutional theory contends that organizations "are influenced by *their institutional environment*, that is, by the norms, beliefs, and cognitions apparent in other organizations."[32] Institutional cultures define appropriate and legitimate organizational forms and activities, constitute identities and interests, and set out the rules and operating principles that guide how organizations within the field should operate. Religions and the secular environmental movement reside in different institutional fields that are organized around different values, rules, and practices. These different, and often conflicting, institutional cultures have kept religions and the environmental movement separated. Religions' institutional cultures and contexts, especially those from Judeo-Christian traditions, also have prevented religions from seeing the environment as an important and religiously salient issue.

My argument draws on the cultural-institutionalist approach of Armstrong insofar as I also emphasize how institutional logics, rules, and cultures shape the goals, strategies, and trajectories of social movements. She argues that new movement organizations arise only during periods of uncertainty and social change, and only in "contexts of creativity" or "situations which place actors in intense interaction with a rich variety of cultural materials to draw from and a high degree of uncertainty about the limits about what is possible."[33] The case of religious environmentalism does not fit neatly into the criteria that allegedly encourage the formation of new movements and new organizations. There are no clear and powerful crises that compelled religious leaders to form a new movement, and cultural resources for religious activists were not more plentiful, richer, or more amenable to reworking during the 1990s and 2000s. Although increased attention to climate change among scientists, the media, and politicians is an important part of the REMOs story, religious activists were slow to recognize climate change as a pressing religious concern. Even when activists identified the important environmental problems they tended to be less attuned to the political dynamics surrounding such issues and more focused on making sense of such problems religiously. Thus explaining how the religious environmental movement emerged poses an intriguing puzzle. The concept of "embeddedness" provides a new way of understanding how, why, and when new movements emerge and develop.

Initially embeddedness referred to "networks of interpersonal relations" and the "structure of the overall network of relations" that direct economic action.[34] Scholars working in organizational and network sociology have developed this concept by specifying different types of embeddedness and showing how each type channels, directs, or limits organizational activity. Four types of embeddedness are particularly important for understanding religious environmentalism. First, structural or relational embeddedness refers to both interpersonal and organizational networks. These are important because membership in particular types of networks can promote trust and foster interdependence, which in turn can lead to joint action. Social-movement scholars have found that preexisting ties among activists can "aid in communication between groups, promote trust, and facilitate the development of a shared ideology and goals that can help reduce the costs associated with collaboration."[35] In addition, activists who are already connected to one another or who have ties (or are embedded) with multiple organizations can help minimize differences in ideology or goals among social movement organizations (SMOs).[36]

Movement organizations often are embedded in a set of preexisting organizational and institutional networks that demand a certain degree of conformity to a shared set of rules, routines, and ordering principles that authorize some types of activity and create a stable institutional world. Organizations that operate in the same institutional field hold one another accountable for following the rules, and when an organization violates any

of these rules, it may undermine its legitimacy, erode trust, and make it difficult to maintain access to resources and opportunities.[37] An organization's
chances for survival and success often depend on conforming to the norm
and expectations of its institutional environment, and those organizations
with multiple, intense, and complex ties within an institutional network
will be more powerfully constrained by the imperative to conform.[38] REMOs
possess strong and often dependent ties to other religious nonprofits, denominations, and congregations, and they depend on organizations within
this field to provide potential supporters, financial resources, and legitimacy. Members of this institutional field promote a logic that prioritizes
distinctly religious goals (for example, saving souls or providing charity for
the poor) over political goals, set the locus of legitimate social action (for
example, people rather than nature), and authorize some relationships (for
example, alliances with groups that share the same beliefs) while discrediting others. In addition, REMOs' connections to denominations and other
religious bodies may make them subject to the authority such bodies wield.
Catholic and mainline Protestant REMOs have formal or informal ties to
their national church and therefore may more closely adhere to the rules
and theological commitments of their national bodies in order to secure
funding and maintain legitimacy than ecumenical or interfaith REMOs.[39] As
will become evident in the following chapters, REMOs' appeals for support,
decisions about entering into alliances, and their decisions about mission,
goals, identities, and protest strategies all hinge on whether or not they will
be seen as legitimate within the religious communities in which they work.

Second, movement organizations are culturally embedded in shared
understandings, values, and beliefs that influence organizations' activities.
Dacin and colleagues argue that organizations within the same institutional
field share "collective understandings that shape organizational strategies
and goals, ideologies that prescribe conceptions of the means and ends
of individual action, and rules systems . . . that categorize organizational
actors and systems of control."[40] Institutional and organizational cultures
constrain and enable SMOs' activities in key ways. They order actors' "understandings of the social world and of themselves by constructing their identities, goals, and aspirations, and by rendering certain issues significant or
salient and others not."[41] And institutional cultures provide the guidelines
or set the limits on the choices SMOs make about framing and mobilization,
alliance formation, and repertoires of contention.[42]

REMOs are embedded in religious cultures that not only made it difficult to engage in environmentalism as the secular movement ramped up
during the 1970s, but also continues to present problems today. There is

no historic green tradition in any religion (except for indigenous religions) in the United States. Apart from the handful of theologians who developed eco-theologies during the 1970s and 1980s (much of which was ignored by religious communities), there were few biblical, theological, or ethical resources available for activists. Religious environmentalism largely was unimaginable thirty years ago. Even today, some religious groups remain suspicious of environmentalism not only because they see it as anti-Christian, but also because they believe it is closely aligned with liberal political causes that run counter to their religious values (for example, pro-choice, pro-gay marriage). This creates barriers to working with secular movement organizations and problems mobilizing support from within their own communities for environmental activism. In addition, some REMOs are tied to systems of religious values and ethical precepts that emphasize the central importance of goals such as promoting individuals' conversion and salvation, encouraging spiritual development, or helping people live a godly life. In religious cultures that stress the primacy of the believer's interior life, there is no obvious place for environmentalism, and REMOs then must convincingly demonstrate that environmentalism is a critical part of the individual's religious or spiritual life. More broadly, most American religions' social ethics emphasize the divine call to alleviate human suffering, and religions historically have addressed such issues as poverty, homelessness, hunger, and health. The environment commonly has not been considered. Thus the broader religious cultures in which REMOs are embedded have made it difficult to make the environment a legitimate and salient issue within the American religious field. Polletta argues that SMOs' strategic choices are filtered through and contingent on the broader cultural understandings that guide activists. She notes that if the "familiar ways of doing things and seeing things" render environmental problems invisible or of secondary importance, then religious activists will be less likely to join the larger movement or mobilize support and participation from religious audiences.[43] For these reasons, REMOs have had to focus their energies on creating green religious traditions and raising awareness about the religious legitimacy of environmentalism within denominations and other religious bodies.

Third, movements are embedded in particular audiences and that they may exercise considerable power over the framing and tactics a movement uses. Polletta notes that strategic choices often conform to the values, ideologies, and practices within a given institutional culture because the risks of nonconformity are too great to entertain.[44] The literatures on embeddedness and social movements have given scant attention to the role audiences play, yet they may be critical for the emergence and development of social movements. Audiences are members of an institutional field (both as organizations and individuals) who have a vested interest in how the institution operates and preserves its culture. Hsu and Hannah claim that audiences often are quite powerful and may be able to prevent new organizations from forming and succeeding because they "control the material and symbolic resources that sustain organizations, their perceptions of whether or not an organization satisfies the applicable codes [i.e., rules] affects their valuations of its worth, and indirectly, the organization's chances of success."[45] Similarly, Zuckerman contends that audiences have the power to confer legitimacy on social actors and their activities, and therefore in order to gain the favor of a particular audience, actors must conform to the audience's expectations, assumptions, and beliefs.[46] In other words, REMOs are embedded in real and potential religious audiences who hold specific views about nature, about what their sacred texts and theological traditions teach about human obligations toward the human and natural worlds, and more generally about the proper role religions should play in the public sphere. As religious nonprofit organizations, REMOs depend on their audience members for financial support, participation, and legitimacy. REMO appeals and campaigns that run counter to their audiences' understandings of religious politics, interpretations of scripture, or application of theological and ethical teachings to the environment face the risk of losing their support. In this way, audiences exercise significant power to constrain the activities of those REMOs with which they are tightly embedded. An important contribution this book makes is to incorporate the analysis of audiences into the study of social movement emergence and development and to show how an organization's embeddedness in audiences influences how they operate.

Finally, REMOs are embedded in particular political arenas in which struggles to control and distribute resources and legitimacy, and to determine the rules of the game, may deeply influence the new movement organization's mobilization and tactical goals.[47] Although REMOs are not oriented primarily toward the state or the formal sphere of public policy and politics, they are strongly connected to and dependent upon religious congregations, denominations, or other types of nonprofits and thus must be attuned to the ways in which authority is ordered and exercised. Religious bodies in which power is centralized pose particular challenges for REMOs because such religions may be able to sanction movement claims that violate official positions on the environment or may require REMOs to adhere publicly to official teachings in order to gain support, legitimacy, or resources. Thus REMO leaders need to know how to frame contentious issues and make appeals for support in ways that will not violate the theological and ethical commitments of their religious partners. They must also be aware of the formal and informal rules and practices that guide intra- and interreligious relationships. For example, REMOs who wish to build an interfaith coalition need to know how to "organize difference" by "enacting rituals or referencing discourses and symbols that are familiar to all group members."[48] REMOs that fail to find common ground among their religiously diverse members or that privilege one specific theological/ethical approach to the environment at the expense of others may struggle to organize protest.

Thus an embeddedness approach allows the analyst to more precisely specify how and why different contexts affect movement emergence, strategic decisions, coalition formation, and outcomes. Because REMOs straddle multiple institutional fields, such an approach can help the analyst identify how the strength of embeddedness affects their efficacy and the degree to which REMOs' connections to multiple organizations may expose them to unmanageable competing pressures, create new opportunities, or severely constrain their activities.

Plan of the Book

In the next four chapters, I develop an embeddedness explanation about social movement emergence, operation, and development. I identify the ways in which different kinds of embeddedness (relational, cultural, and institutional) cluster to shape the choices activists make as they establish a new religious movement. Chapter 2 unpacks the founding stories of religious environmental organizations to show how activists' location within particular religious settings (from churches to denominations to interfaith parachurch groups), religious traditions, and particular ecological settings work together to create several different pathways to the foundings of REMOs. Chapter 3 focuses on the myriad choices activists face as they attempt to establish new movement organizations and how those choices are limited by the religious context in which they work. I describe how a REMO's embeddedness in particular theological, biblical, and organizational cultures shapes how it crafts its mission and goals. I then demonstrate how organizational mission influences choices about organizational form and the selection of ecological issues, protest tactics, and mobilization strategies. Chapter 4 charts the creation of green religious traditions and, more specifically, how activists borrow and adapt images, theological warrants, and ritual practices from their own traditions to craft new environmental ethics. They do so constrained not just by what is available in the tool kit, but also by how such borrowing and adapting might be perceived by potential constituencies. I show how being embedded in particular religious traditions and real and imaginary audiences shapes the formation of the new culture of religious environmentalism. Chapter 5 describes the interorganizational networks among REMOs and between REMOs and other organizations. I explain how ties to specific national bodies and religious communities and specific theological commitments encourage cooperative partnerships between like-minded REMOs but discourage alliances between REMOs from different traditions and between REMOs and secular movement organizations. Chapter 6 reviews the major tenets of an embedded perspective on movement emergence and the primary findings about the emergence and development of the religious environmental movement. I then identify several lessons about religious social movements that emerged from the research and offer an explanation about why religious social movements do not neatly follow existing theories about social movement emergence, mobilization, and outcomes. Finally, I conclude with a discussion of the relationship between religious and nonreligious environmental movements, the ways in which the former have influenced the latter, and what the future holds for religious environmentalism.

TWO

The Emergence of the Religious Environmental Movement

I live on a farm. We raise lamb and beef—no growth hormones, no antibiotics. We try to direct market all of it. There's a reason why I'm telling you this; I was a vegetarian for many, many years, and it was in response to how we raised animals and the fact that we were eating at the expense of other animals. So for me, living my life has a lot to do with living my faith. There's no point saying that I am a "so-and-so" if you don't live what you say you believe in. I'm very deeply connected to the world I live in, and it was just becoming clearer and clearer to me and my husband who also worked on issues that deal with energy and climate that something was not right, okay? And then realizing that, well, if I know something is not right then I have an obligation to do something about it.

This is Renee Gopal's initial answer to my question about how the Prairie Climate Stewardship Network got started. In 2006, she was serving as the executive secretary to the North Dakota Conference of Churches when she brought the issue of climate change to the attention of the denominational leaders who sit on its board. Her plan was to have the conference issue a statement on global warming that was tailored to the needs and sensibilities of North Dakotans, and then have the conference suggest how churches could engage in practical climate stewardship. She was confident the denominational leaders would support her proposal because it was not controversial or divisive but instead was grounded in shared values and the biblical call to be good stewards. She was not prepared for the resistance she met:

The way I phrased this thing is it would be an act of hope and faith, and it's going to be based on the shared theology. Given that we do have this shared theology of creation, it seemed to me that we could speak with one voice. It

wasn't as though we were talking about pro-choice/pro-life, same-sex mar-
riages, you know; it was a very different sort of playing field. Another reason
why I thought the conference would be a right place for this was that it has
a particular committee that focuses on rural life. And it was very clear to me
that if we are serious about rural life ministry in North Dakota, we need to be
addressing climate stewardship and global warming. . . . In the end, when the
statement was brought to the leaders after it went through several rounds of
feedback and editing, eight of the thirteen agreed to the statement. One was
dubious that human activity was contributing to climate change; one said
he was not going to sign it; and one said he would have to bring it to his full
synod. One member was very exasperated and said, "Well, I guess what we're
going to have to do is affirm each other's statements." So when I heard that I
thought, "Wow, this isn't going to happen." So it became clear to me that another means had to be found to engage people of faith in climate stewardship.

Renee soon quit her job with the conference and quickly learned that
there were no religious groups working on climate change or other environmental issues in the state. Demoralized but undeterred, she used her knowledge of North Dakota churches to pull together a group of like-minded lay
and ordained Christians from a variety of traditions to create the Prairie
Climate Stewardship Network (PCSN). The group decided it should provide
faith communities with resources about the science of climate change, models of how to reduce their carbon footprint, as well as teachings from various
traditions about the relationship between faith and the environment. They
hoped all of these resources would persuade Christians that it was not only
okay but also necessary for them to be environmentalists. And they needed
to "speak to North Dakota in a way that conservative farmers and ranchers
would understand." The organization started with $750, and for more than
a year Renee worked without a salary. She collected information, created a
web page replete with a carbon footprint calculator and the latest scientific
studies about the effects of climate change on the state, and a document
for churches titled "Global Warming and Creation Care." Eventually she
raised enough money through private donations (denominations were not
interested or able to help she reported) so that PCSN was able to hire Renee
full time.

When I asked her about how she made it through the first year of the
REMO's life, Renee replied with great passion:

If you really say you love God, how can you not act out of love for God? If you
really say you love God, what does that mean? For me it means an abandon-

ment of yourself before your God, to do God's will. . . . The door is open; you walk the path; you know there is no other choice, because this was the choice in your heart and you do it. One day I was driving to one of my meetings, and I was so overwhelmed by everything, and then it just became very clear for me—"anything you [God] ask me to do I will do for you because you will make it possible." So it's more than just environmentalism; it's about a relationship, a very deep relationship.

The account of the Prairie Climate Stewardship Network highlights a set of common biographical, eco-political, organizational, and cultural influences that shape the emergence of the new social movement organizations. First, many founders of REMOs have a profound and long-standing interest professionally or personally in environmental issues. Some, such as Ms. Gopal, spoke about their deep connections to the land; others felt compelled to yoke their faith and professional lives more tightly; and a significant number decided to establish a REMO after having some kind of awakening in which a spiritual experience of the natural world sensitized them to climate change or pollution. For many activists, forming a REMO is not just about trying to alleviate some environmental problem; it is a response to more fully and authentically live a religious life. Second, activists use ties to a variety of religious organizations (denominations, congregations, parachurch groups) to mobilize resources and supporters. Sometimes organizers like Ms. Gopal fail to win the support of official religious bodies and thus are forced to find or create alternative religious networks from which to draw supporters; in other cases religious organizations such as the National Council of Churches will sponsor the creation of a new REMO. Third, REMOs' emergence is fueled by a specific religious culture as founders draw on particular beliefs about the sacredness of creation, stewardship, or justice to recognize and construct opportunities for activism. Finally, REMOs emerge in particular eco-political contexts that shape choices about the mission, strategies, and alliances. In the case of the PCSN, the conservative political and religious setting of North Dakota and the growth of the energy industry in the state led Ms. Gopal to emphasize climate change and alternative energy as critical issues, and to focus resources on more practical matters of energy conservation. More importantly, REMOs must be attuned to the religious interests and sensibilities of the audiences in particular settings. In the case of PCSN, they framed appeals around stewardship (both in terms of the biblical warrant and as a matter of resource management) rather than the more controversial religious themes of justice and spirituality since these would be less likely to appeal to the state's conservative Christian population.

In short, the emergence of the religious environmental movement can best be explained by understanding how activists' embeddedness in particular organizational, cultural, and eco-political contexts shapes their choices about organizational form, frames, tactics, and resource acquisition. Diverse types and combinations of embeddedness shape how and when REMOs emerged. This focus on embeddedness moves the explanation of movement emergence away from the overly structural and macroemphasis of the political opportunity structure perspective and toward a novel perspective that emphasizes the concrete dynamics of emergence. In the next section of the chapter, I outline the existing explanations of movement emergence and then draw upon recent work about the emergence of new organizational forms and calls for a more dynamic, relational approach to the study of social movements to develop my explanation for the rise of the religious environmental movement.

Explaining Movement Emergence

The Canonical Account

The most common explanations for movement emergence combine key concepts from political opportunity structure (POS) theory with those from the resource mobilization perspective.[1] Most include four broad sets of factors that encourage movement emergence: social/political/economic changes that create and/or deepen social problems; a political system that is open to challenge or that has been affected by critical events such that there are now new opportunities to push for social and political change; insurgents who have some set of grievances about the social order and can articulate persuasive demands that authorities take remedial action; and pre-existing organizations and/or advocacy networks that provide insurgents with the necessary resources to push effectively for change and get the movement up and running. According to three leading scholars in the field, "Most political movements and revolutions are set in motion by social changes that render the established political order more vulnerable or receptive to challenge. But these 'political opportunities' are but a necessary prerequisite to action. In the absence of sufficient organization—whether formal or informal—such opportunities are not likely to be seized. Finally, mediating between the structural requirements of opportunity and organization is the emergent meaning and definition—or frame—shared by the adherents of the burgeoning movement."[2] In other words, movements emerge when the political system is rendered vulnerable by a significant social change or crisis; when insurgents are able to construct a set of shared grievances, solutions, and collective identity; and when they are able to create strong and effective organizations to push for redress.

While this is the canonical account, a number of scholars have voiced serious concerns about its theoretical imprecision, strong structural determinism, and lack of attention to the role of culture, agency, and internal movement dynamics. Polletta notes that "without theorizing the actual dynamics by which movements emerge, sociological accounts of movements can fall back into barely disguised structuralism (political turmoil generates movements, which in turn generate cultural challenge)."[3] Goldstone summarizes the extant critiques in the following way:

> POS has three major difficulties. (1) In pointing us to the "political," it emphasizes conditions relating to states, tending to neglect the role of countermovements, allied movements, critical economic conditions, global trends and conjunctures, and various publics. (2) In pointing to "opportunity," as the label for changes relevant to movement actions, it tends to neglect how, in many cases, adversity—such as threats, excessive repression, or countermovement actions—can energize and elevate movements, increasing their support and chances of success. (3) In pointing to "structures" . . . it tends to emphasize pervasive large-scale conditions and suggest necessary and sufficient conditions for certain outcomes. In fact, different groups may face very different group- and issue-specific conditions regarding their mobilization and success, and such conditions are often more fluid and relational than they are "structural" in character.[4]

In particular Goldstone urges scholars to look closely at the set of complex relationships movements have with various political institutions and other collectivities in order to understand how those relationships encourage or discourage the formation of new movements. Similarly, Jasper advocates for an explanation that stresses the ways in which insurgents who are embedded in particular arenas ("sets of resources and rules that channel contention into certain kinds of actions") with a particular set of "players," come to make choices about goals, tactics, and framings as they interact with one another:[5] "A social movement comes to life as various individuals, formal organizations, and informal networks and groupings coordinate events and share some goals and know-how about tactics."[6]

The rise of religious environmentalism raises two additional problems with the dominant explanations of movement emergence. As evident in the opening vignette in chapter 1, REMOs are primarily oriented toward

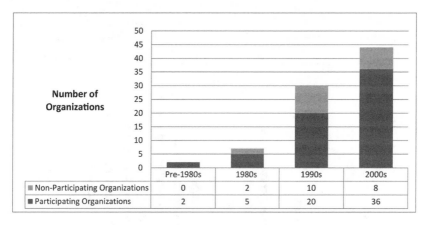

Number of Organizations	Pre-1980s	1980s	1990s	2000s
▣ Non-Participating Organizations	0	2	10	8
▪ Participating Organizations	2	5	20	36

Fig. 2.1. Foundings of all REMOs in the United States

religious communities and individuals rather than the state or other distinctly political actors. Although they engage in some political advocacy, they rarely direct their challenges to the state. They are not fundamentally interested in seeking political solutions to environmental problems or to reform the political sphere, and they are rarely interested in building alliances with politicians, parties, or political organizations. They act with little regard for shifts in leadership and power at the state or federal levels and thus are unlikely to be influenced by political allies (or the lack thereof) or the possibility of state repression of their activities. The movement seems to have emerged with little awareness of new political opportunities or new vulnerabilities in the political system that they could exploit.

In addition, the twenty+ year time period in which the new movement emerged poses explanatory problems for the political opportunity perspective. Recall that the contemporary, politicized environmental movement got started with Earth Day in 1970. It was particularly active and successful during the 1970s and 1980s when many of the landmark environmental laws were enacted and the movement deepened its base and developed its full repertoire of contention. Yet very few REMOs were founded during this period, and it really was not until the late 1990s and early 2000s that REMO foundings took off, as seen in figures 2.1 and 2.2. Thus the movement emerged over an extended time period in which control of the presidency, Congress, and state governments changed hands regularly, the economy oscillated between booms and busts, and the United States engaged in a decades-long war against terrorism. However, there is little relationship between these external changes, crises, and contexts and the founding of REMOs. There is

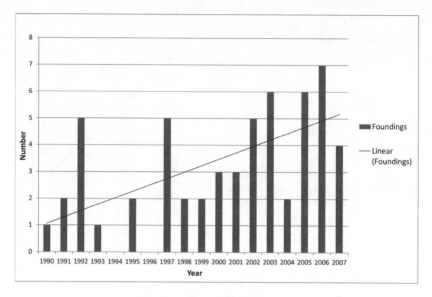

Fig. 2.2. Foundings of participating REMOs since 1990

no obvious and consistent pattern in which the timing of REMO foundings lines up with specific political opportunities or macrolevel social changes and crises. Yet macroevents, such as the increasing concern with and scientific evidence for climate change, are important for the emergence of the new movement—just not as direct catalysts as predicted by POS theory. Instead, large-scale social and political changes may be important by changing the "normative expectations and cognitive frames around how to conceive of and address social ills."[7] That is, large-scale structural changes and events can shape movement emergence by changing how activists think about social problems, how they rank order problems, and how they imagine novel solutions.

Most REMOs started out with limited access to resources, participants, and institutional legitimacy within the sphere of religion. Many began with just a few individuals and pocket change, as in the example of the Prairie Climate Stewardship Network, and only a handful acquired the kind of financial resources and supporters to launch large-scale and sustained actions. REMOs largely relied on volunteers during the early years of the organizations, and few garnered the support of congregations, judicatories, or national denominations. They were not enmeshed in dense interorganizational webs that provided access to social-organizational, human, or material resources, as in the case of many other successful movements (for example, the peace or civil rights movements).[8] Although REMOs started out like many other social movement organizations—small, local, mostly volunteer, and resource poor—they were not embedded in resource-rich environments nor were their causes considered legitimate and worthy by many of their potential constituents. As a result, most continue to struggle to survive.[9] Even after weathering the difficult early years, most REMOs have small paid staffs (usually between one and three staff members), budgets under $100,000, and small pools of active supporters and members. In short, REMOs were founded in a less than friendly religious institutional context, in an ever-changing political environment that was no more or less open to their grievances, and in dozens of ecological settings in which there was no clear environmental disaster or threat to catalyze protest. Their existence seems to defy the predictions of the dominant explanation in the literature. So how did this new movement emerge?

The Embeddedness Explanation

In this section I outline an alternative explanation of movement emergence that follows recent work that calls for greater attention to the concrete processes of emergence, network relations, and the dynamics within multi-institutional fields, and the role that strategic choices of insurgents play in SMO foundings.[10] The emergence stories from my study organizations are as varied and colorful as the organizations and founders themselves, but they share several common features. First, most of the nascent REMOs reported that some event or experience prompted the founders to begin their organizing efforts. For many, the catalyst made an individual aware of his or her call to take environmental action as a person of faith and was interpreted as an awakening or conversion type of experience. Unlike the sort of large-scale structural catalysts identified as crucial in the literature (for example, recessions), the catalysts for REMOs were more modest and local, such as ongoing problems with toxic waste from brownfields in New Jersey, attending an environmental conference, or having a spiritual experience while backpacking.

Second, REMO leaders were culturally embedded in specific theological worlds and organizationally in particular religious networks. The former would prove important because they provided founders with the language, images, and values they could apply to the environment and use to connect with the religious sensibilities of their potential constituencies. The latter type of embeddedness provided them with a deep understanding of how religious bodies (from congregations to national denominations) operated and important contacts through which they could try to mobilize various resources.

Third, the founders all faced the same set of organizational challenges as they endeavored to create a new type of religious organization. As "institutional entrepreneurs," the creators of REMOs had to gather resources, identify opportunities, and persuade others to act with them to capitalize on the new opportunities, gain legitimacy for religious environmentalism by developing a justification for the new movement, and integrate REMO goals, values, and practices within the operative institutional orders of American religion and politics.[11] These challenges were heightened for the organizers of REMOs because environmentalism and direct politics were considered illegitimate or at least not very important to many religious bodies and individuals. Moreover, much of American Christianity was immersed in battles over cultural issues (that is, abortion and homosexuality) and/or busy with a host of human-focused benevolent activities and thus had little interest or energy for adding another "cause" to the long list of issues on which congregations and denominations worked. It was no small feat for the founders of REMOs to get established and succeed in this environment, and it was largely due to their entrepreneurial skill and strategic choices.

However, the emergence of REMOs is not simply a story about the conjuncture of catalysts, biographies, and entrepreneurship. It is a story about how REMOs' embeddedness in particular religious cultures and organizational networks, audience, and eco-spaces led their founders to interpret specific events as religiously significant, act on environmental interests, or sell their vision for faith-based environmentalism to fellow religionists. A founder's embeddedness in a given tradition, for example, may provide her with interpretive resources out of which she may perceive or construct environmental needs and opportunities. Armstrong argues that "features of the cultural milieu in which actors are embedded influence the likelihood that they will mobilize available resources to create new organizational forms."[12] Moreover, cultural embeddedness provides institutional entrepreneurs with the beliefs, values, and ideas they use to formulate new frames that are critical for gaining legitimacy and other resources.[13] Similarly, one's location within a specific ecological-political space may alert one to the most salient issues and how best to frame the work of the REMO. This is evident in Renee Gopal's focus on energy and climate change, as well as her concern that she not alienate the rural conservatives who make up her potential pool of supporters. Attentiveness to and understanding of a REMO's place within the ecological and political setting is crucial if the new movement organization is to identify and then mobilize its intended audience.[14] Accordingly, REMOs also are embedded in real and imagined audiences who not only may one day become the rank-and-file members of the organization and financial supporters, but also are important granters of legitimacy. REMO founders could not ignore the environmental beliefs and faith commitments of their intended audience as they went about the tasks of mobilizing resources, developing frames, and establishing their missions. Finally, REMOs are embedded in multiorganizational fields comprised of other REMOs, secular environmental movement organizations, a variety of religious bodies, and alternative energy companies, which create more or less hospitable environments for emergence. These interorganizational network relations were critical sources of resources and models of organizing and knowledge, but they also constrained or facilitated certain courses of action. In sum, the nature and depth of REMOs' embeddedness in particular cultural, ecological, and organizational contexts directed how they emerged and developed.

Elements of Emergence

Biography

Most founders were biographically predisposed to get involved in religious environmentalism. Nearly 30 percent were already working for a religious nonprofit doing advocacy and organizing around social justice issues. About 40 percent were clergy, and most of the interviewees noted that the founder of their REMO had a deep and long-standing interest in the environment. A handful sought to merge their professional work as environmental scientists or advocates with their faith lives or to turn their ministerial call into one focused on healing the natural world. For example, Chris Bright, founder of Earth Sangha, worked as an analyst at an environmental think tank and was a practicing Buddhist. He and his wife were interested in starting a sangha, or community, but wanted to do more than focus on sitting or meditating, which is the most common religious practice done by members of a sangha. He noted that "instead of just sitting, which is very important, we hoped the organization would have some sort of environmental work that would yield a real tangible benefit." They started Earth Sangha in 1997 with the goal of reforesting the mid-Atlantic and other parts of the world first by establishing seed banks and nurseries of local, native species of plants and then replanting in damaged areas around Washington, DC, and the Dominican Republic. Similarly, Rev. Peter Sawtell's educational background in environmental biology and religion led him first to parish ministry in the United Church of Christ and then toward founding Eco-Justice Ministries:

> My college major was essentially in environmental biology, so a very deep understanding of that perspective is second nature to me. That was combined with a very significant and more in-depth sense of social justice issues that I picked up in seminary, in terms of institutional racism. So those are long-standing perspectives for me that I brought into the work that I did in parish ministry. And the late '80s and early '90s, I did doctoral studies in the field of religion and social change . . . and then really the trigger point in all of that was during the summer of 1994, I settled down on a vacation week to read through some of the classic and some of the emerging pieces on the environmental crisis. And that really was a profound conversion moment for me, as the stuff that I knew in my head came across with clarity that had a very significant emotional and ethical impact. . . . It took a few years to figure out how I might engage in that, especially in a religious setting. Starting in 1997, I took on the volunteer role of eco-justice coordinator for the Rocky Mountain Conference of the United Church of Christ, which was a National Council of Churches–related program. Over a few years I became very aware that 10–15 hours a week wasn't going to make the difference that needed to be made. So early on, in 2000, I was at a point of vocational shifting, and said, "If I'm really going to care about this, I need to do it full time." That was the point at which I started the agency.

In short, religious and environmental education, years spent camping and hiking, a reverence for the natural world, or a job in the environmental field predisposed individuals toward starting REMOs. Yet biography alone was not sufficient, as evinced in these accounts. Something had to happen to push the founders toward acting on these predispositions. Like Reverend Sawtell, nearly one-quarter described some kind of conversion, awakening, or an "aha" moment that turned them toward organizing the new movement.

Catalysts

Interviewees described a variety of religious and environmental catalysts or triggers. Sixty percent of the triggers were religious in nature, and they ranged from individual encounters with the sacred to judicatory or denominational actions, while the rest were rooted in some environmental problem or the result of a green organization prodding religious groups to get involved. For some, seeing ecological disasters such as oils spills (for the founder of Earth Ministry) prompted them to take action, while for others the catalyst was when the sacred was revealed in nature. For example, Nigel Savage, founder and executive director of Hazon, a pan-Jewish organization, described two moments when he experienced what Jewish theologian Abraham Joshua Heschel calls "radical amazement":

> It was in Golders Hill Park, in Northwest London, on a cold clear early February morning. I was walking through the park, and I was reading to myself an affirmation from a book by Louise Hay. The words were: "I listen to the divine and rejoice at all that I can hear." And as I walked through the park, and said these words, and looked at the trees, and felt the cold, and saw the ducks on the lake, I felt my relationship with the world around me change.

The second moment came at the end of a four-day hike from the Mediterranean Sea to the Sea of Galilee. The group finished the hike on the Sabbath "utterly exhausted and utterly exhilarated," and they walked into the water to celebrate the ritual of Havdalah (prayers that mark the end of the Sabbath and start of the new week):

> And even though I have made Havdalah countless times before in my life, this was a different Havdalah, just as davenning [praying] had been different for me during the hike. The Jewish people didn't arise in a synagogue or temple; our prayer life didn't develop in a suburban building; the words of our prayers and of the psalms did not begin with a siddur. The beauty and rhythm and wisdom of Jewish tradition arose in an encounter with the majesty and with the *awe* of the physical world around us. Something about being outdoors, being exposed, being physically challenged, being bereft of electrical toys and the protection of metal and brick and glass, being in contact with the wind and trees and small animals; all this had a profound impact on me.[15]

The experience of the hike and that ending ritual would become the trigger to quit his job in the financial industry and devote himself to developing an environmentally active Jewish community. Other activists spoke about their encounters with environmental disasters like mountaintop removal in Appalachia (Christians for the Mountain and LEAF), toxic waste (Jesus People against Pollution), or poverty and ecological degradation in Latin America caused by the US capitalist-driven lifestyle (Eco-Justice Collaborative). These encounters were seen through the lens of an individual's religion and coded as a conversion or call from the divine to act. An important facet of these awakening catalysts is that they are viewed as a call to live out one's religious commitments authentically. This is seen in Renee Gopal's discussion of her sense that she had no choice but to start PCSN once the North Dakota Conference of Churches reached a stalemate on the climate change statement. The two young founders of Faiths United for Sustainable Energy (FUSE) reported that they grew up as Reform Jews in families and congregations active in a host of social issues and were taught that being involved was a "moral necessity." Just as the civil rights movement was the moral necessity of their parents, they saw climate change as the issue that demanded their involvement as observant Jews. The notion of authenticity or that environmentalism is part and parcel of what it means to be a religious person or community is a defining feature of the new movement.

A number of founders were moved to establish their organization as a result of proenvironmental work by congregations, judicatories, or national religious bodies. For example, Presbyterians for Restoring Creation emerged five years after the General Assembly of the Presbyterian Church (USA) adopted a policy report ("Restoring Creation for Ecology and Justice") since little had been practically accomplished to fulfill the mandate to incorporate care for creation into the life of the denomination. The New Community Project (NCP) was founded after the Church of the Brethren disbanded its office for environmental programs. Founder David Radcliff, who had run these programs, noted that "there was a general outcry in the church and some interest in keeping me engaged in some of these things. So out of that sort of groundswell of concern and interest we kind of gave birth to the New Community Project."

Some REMOs were established because some existing environmental organization offered funding or even created REMOs themselves, such as when the NCC created several state-level global climate change campaigns. Such movement midwife organizations often pulled religious leaders together and urged them to create their own particular REMOs. For example, the NRPE's organizing work led to the creation of the Coalition on the Environment and Jewish Life (COEJL) and the Catholic Coalition on Climate Change. Secular movement organizations also prompted religious activists to form new groups. The Interreligious Eco-Justice Network (IREJN) was formed after a group of Connecticut clergy, who had participated in a Clean Water Action air pollution campaign, engaged in a prolonged conversation about how religions could be involved in working for environmental justice. Similarly, GreenFaith was founded after "a couple of people, religious leaders—lay and ordained—from different parts of New Jersey had gone to the Rio Earth Summit and had been really moved by that experience. They came out of that feeling it was important for there to be an organization in New Jersey to help the state's religious community engage in environmental issues from a religious perspective."

In short, the founders of REMOs felt called to create these new organizations through a variety of religious events and experiences as well as by the efforts of secular actors who urged religions to join the environmental movement. Some REMOs were started in response to formal denominational or congregational environmental initiatives, while others emerged because religious bodies did not act. Many interpreted the growing problem of global warming or ecological problems such as mountaintop removal as religious problems that required them to act. How the founders of these new organizations came to interpret these events and experiences as a divine or sacred call depended on how they were embedded in particular religious cultures, interpersonal and interorganizational relationships, and specific eco-political spaces.

The Role of Embeddedness in the Emergence of the New Movement

As noted earlier in the chapter, being embedded in specific religious cultures provided founders with a set of tools and a worldview that helped them make sense of their experience of the natural world. The founders of the new movement were deeply religious individuals who were steeped in particular traditions' sacred stories and hermeneutics, moral principles and ethical dictates, rituals and prayers, and teachings about justice and stewardship. Jonathan Merritt, who established the Southern Baptist Climate Initiative, noted how his grounding in scripture, evangelical Biblicism, and reverence for the Bible, and the more general Christian tradition of seeing God as an active agent in human affairs, triggered his interest in environmental activism:

> Well, it started a number of years ago. I was sitting in a systematic theology class in Southeastern Seminary in Wake Forest, North Carolina. . . . We were talking about revelations, and he [the professor] said, there are two primary forms of the revelation of God. There's sort of a special revelation that we get through the scripture that can lead you to salvation; and then there's the general revelation of nature, and scripture speaks about that multiple times. And then he said something that was very interesting . . . there are two books that God has written. So when we destroy God's creation, which is God's revelation, it's similar to tearing a page out of the Bible. And the way that he said that, it really reached me because I thought, "Man, I would never tear a page

out of the Bible." But it's sort of the same thing where we're defacing God's revelation. So that really was a shifting perspective for me because, up until that point, I was the opposite of an environmentalist.

Merritt would go on to work with a number of Southern Baptist seminary professors and church leaders to draft a call for his denomination to address climate change in concrete ways. As will become evident in chapter 3, Merritt succeeded in launching this initiative because he stayed firmly within the boundaries of Baptist theology and denominational politics.[16]

Similarly, David Radcliff, founder and executive director of the New Community Project, discussed how his ties to the Church of the Brethren, and more broadly the Anabaptist traditions, shaped how he developed the mission of the new REMO:

> Jesus is a role model or person who invites us to follow, which He very clearly does, of course, in the New Testament. And so our church has, I think, been a little more gospel centered. . . . One of my favorite hymns growing up, and I was just thinking about this the other day, was, I don't know what the title is, maybe it's "Where He Leads I'll Follow"—that's the refrain—follow Jesus every day. . . . This idea that Jesus is there as someone to show us the way literally, in terms of kind of modeling himself and also inviting us to come and follow his example, and we take it very seriously. The Sermon on the Mount, for instance, if we had to look at our hermeneutic, it would be the Bible, of course, but then also, we ask, and the Church of the Brethren has said the New Testament is our creed.

The New Community Project stresses orthopraxy or ethics over orthodoxy (right belief) and takes seriously Jesus's teachings about the poor, and in doing so conforms to the key tenet of the Anabaptist tradition.

REMOs' founders also were enmeshed in a variety of religious networks that not only shaped their understanding of the relationship between their tradition and the environment, but also how they went about organizing new movement organizations. In some cases, these networks even created the opportunities for the fledgling activists to get started.[17] Much of the work on networks and social movements has demonstrated how important networks are for the acquisition of resources and for mobilizing individuals, but recent work has begun to show how movement networks may operate as crucibles of meaning construction. Diani argues that "movement networks are not a mere practical channel for the circulation of essential resources for action. They are also networks for the circulation of meaning, i.e., social systems where conflictual cultures are elaborated, specific solidarities develop, and thus where political activities and cultural practices are assigned a shared meaning."[18] Similarly, Campbell notes that networks may play an important role in helping activists come to some degree of consensus about problems, solutions, and goals.[19] For the nascent religious environmental movement, interorganizational networks were important incubators for ideas and models of action. For example, Aleph's strong connections to the Jewish Renewal Movement shaped how it approached its project on sacred foods as it explicitly drew inspiration from the founder's ideas about eco-kashruth.[20] COEJL's founders' strong ties to nearly thirty other Jewish organizations helped them develop its mission and establish its legitimacy.

Founders' ties to clergy as well as to similar types of religious groups also shaped the emergence of REMOs by influencing the process of starting up new organizations. Sometimes REMOs emerged when members of extant clergy networks gathered for conversations after attending a conference together or participated in activities sponsored by a secular environmental group, as in the case of Faith in Place (FIP). This REMO emerged as the result of a program initiated by a staffer (the Reverend Clare Butterfield) at Chicago's Center for Neighborhood Technology. The original program gathered religious leaders from across the Chicago region in 1999 to discuss, pray, and act on issues of environmental sustainability. The group learned that sustainable food and energy were the two common issues that generated consensus, and by 2003, they decided to break away from the Center for Neighborhood Technology and form an independent religious organization in order to focus on helping religious communities understand that they have a moral and religious obligation to work on behalf of the environment.

Other REMOs emerged when someone activated broader networks of congregations, religious advocacy organizations, and/or denominational officials to organize a local or regional conference on a particular ecological problem. Such efforts served to identify issues and promote an interest in doing more than talking together. For example, Congregations Caring for Creation (C3), a Minnesota REMO, emerged from the joint work of a variety of Minneapolis–Saint Paul congregations, the Lutheran Coalition for Public Policy, and the secular Alliance for Sustainability. After attending a conference sponsored by the North American Coalition for Christianity and Ecology in 2004, the Minnesota contingent continued to meet in order to promote sustainability within the state's religious communities. The C3 website described the founding in this way.

Over 40 clergy and lay persons from Catholic, Protestant and Jewish congregations and leaders of environmental groups began meeting to design ways in
which local congregations can work with their local communities to achieve
tangible improvement in ecological health and sustainability of their biore-
gions. From those conversations, Congregations Caring for Creation became
a joint project of the Alliance for Sustainability and the Lutheran Coalition
for Public Policy.[21]

The Baltimore Jewish Environmental Network (BJEN) emerged in a
similar way after participants at an all-day conference on how Jewish groups
can promote sustainability decided to move beyond conversation. Its
founder, Rabbi Nina Cardin, noted, "After the closing session of the conference, we had a little town meeting and said anybody who's interested and
wants to help us continue the conversation and create some sort of structure
to continue the conversation let us know. Well, about half the people there
(about 60) were asking us to keep talking, and it was out of that that the
Baltimore Jewish Environmental Network was born." As the C3 and BJEN
examples illustrate, founders relied on their ties to both religious and secular
groups to get their fledging REMOs' foundings off the ground.

Finally, some REMOs were embedded in specific eco-spaces that directed or limited how they went about establishing their new organization.
Many REMOs were created with an eye toward addressing local or regional
environmental problems. As such, their founders were well aware of not
only what issues were important and could gain traction among their potential constituencies but also of the political terrain in which they would
have to navigate. Founders were particularly concerned with tapping into
the religious ideas held by audiences within specific eco-spaces and drawing on these meanings as they framed the problem and solution. This was
especially the case where the REMO faced a conservative religious and/or
political audience, as illustrated in the opening vignette about the Prairie
Climate Stewardship Network. Similarly, both Christians for the Mountains
and LEAF, located in the coal country of Appalachia, faced a skeptical and
even hostile audience, and both organizations took great care framing the
issue in the early days. Pat Hudson of LEAF noted that they intentionally
used the term "creation care" to signal their distinctly Christian but not
overly liberal orientation:

We came up with a packet, and on the cover of the packet, it says, "LEAF is a
Christian fellowship of Tennesseans whose faith leads them to take actions for
Tennessee's environment. Concern for God's creation is not a matter of being

liberal or conservative, republican or democrat. We believe people of faith can look beyond such distinctions and do the Lord's work together." So right up front, we're saying, "Don't throw the message out because it's not coming from your particular denomination or your particular perspective—give it a chance." . . . So we were careful in how we crafted it. We realized that for many of the conservative congregations that we hoped would at least listen to the message, the word "environmental" was a little bit of a red flag. And it's been made that way, unfortunately, because of these stereotypes, because of the sound bytes, because of the entrenchment of both sides over the past several presidencies, and we just wanted to get beyond that. We began to use the words "creation care" because that is a term that very much resonates with people who, when they hear the word environmental or environmentalist, they tend to think of a whole package of "liberal issues." . . . What we found was people are perhaps a little bit suspicious initially, but we are very upfront and say, "Take this packet of information back to your congregation, and use it any way you want to. It is up to you." We're not going into a Church of God and telling them this is how you have to use it, or this is what you have to believe. We're saying we would like to bring this issue to your attention; we think it's an issue that is of importance to all east Tennesseans; and most of them don't know that mountaintop removal is happening. Most of them have not paused to think about the connection between environment and faith. It's really been off the radar for a very long time. . . . And then when you say "now take it and use it in your faith tradition, in whatever way your faith calls you to do it," it really takes the pressure off. And we found it works, it works really well—sometimes not as quickly as we would like because it has to percolate.

In this extended answer to my question about how LEAF managed to get established in a conservative area, Ms. Hudson identified her understanding of the kind of resistance she expected to encounter because of the liberal stigma attached to environmentalism, the fiercely independent congregationalism of the region, and the unique ecological assets and threats in east Tennessee. By using terms associated with conservative Protestantism (that is, "creation care"), grounding their appeal in the theological traditions of the Reformation, and adopting a nonconfrontational, low-pressure appeal, she hoped to win over conservative Christians.

A number of interviewees spoke about the political landscape in which they were trying to get established and how they had to tread carefully lest they alienate potential supporters or collaborators and lose their fledgling legitimacy. For example, Voices for Earth Justice (VEJ) actively sought out partnerships with unions (steelworkers, autoworkers, electricians) in order to appeal to an important audience in the heavily unionized state of Michigan. Alan Johnston of Christians for the Mountains (CFM) spoke about the challenges of mobilizing protest in the coal fields during a down economy. One member of the group received death threats after she led an effort by CFM to win a court injunction to stop the opening of a new mine, while the coal companies urged religions to "do the work of the church, getting people ready for heaven, and let business run business." When REMOs were embedded in these kinds of political contexts, the founders carefully framed their appeals in nonantagonistic ways by using religious language and ideas that were widely shared, and they tried to create programs that would appeal to the interests of their potential supporters (for example, saving the mountains and empowering Appalachians in the case of CFM and forming a green-blue alliance in Michigan in the case of VEJ).

Thus REMOs emerged as different elements combined or converged in particular places and at particular times. While activists' biographies predisposed them toward creating the movement, how they were pushed or persuaded to start up an organization depended a great deal on the type and depth of their embeddedness in specific religious cultures, networks, and eco-political spaces. In the next section of the chapter, I rely on four case studies to illustrate the different pathways and how embeddedness influenced the formation of REMOs.

Pathways of Emergence

The examples described in the preceding section and more generally the accounts provided by the interviewees suggest that REMOs emerged along several different pathways with a host of minor variations within each (see table 2.1). All of the pathways depended on or emphasized one or more types of embeddedness. One path was organized by cultural embeddedness as REMO founders' particular religious traditions helped them recognize the connections between faith and the environment, their religious duty to act, and how best to mobilize their constituency. In a second pathway, the founders' network or relational embeddedness provided opportunities, resources, and/or the impetus to organize. A third pathway illustrates how tradition and networks worked together in the creation of a REMO. As will be evident in the forthcoming analysis, the three pathways overlap, but they are distinct insofar as each emphasizes a different type of embeddedness. The final pathway is restricted to already existing religious

2.1 Pathways of emergence

Pathway	Critical Elements	% of All REMOs	Example
Cultural Embeddedness	Culture provides tools for mobilization	20	Restoring Eden
Relational Embeddedness	Networks provide resources, opportunities, and constraints	27	National Religious Partnership for Environment
Culture and Networks	Use traditions to sensitize and mobilize local networks	32	Partnership for Earth Spirituality
Mission Expansion	Staff/Board member as driver; frame extension	21	Progressive Christians Uniting

nonprofit organizations and reflects the expansion of an organization's mission and programs.

Enmeshed in Tradition

All REMOs are embedded in a particular religion tradition or set of traditions (in the case of interfaith organizations). The doctrines, values, and rituals within a tradition provide REMO founders with a set of tools they use to understand the relationship between religion and the environment and even come to see what counts as an ecological problem. For example, most Christian traditions recognize God's concern for the poor and direct believers to help the poor by providing charity or by changing the practices and structures that cause poverty. However, few traditions claim nature requires the same attention. Thus saving an endangered species, for example, may not even register as a problem, much less as a moral imperative. The tradition not only alerts the founders to what kinds of issues are religiously legitimate, but also what kinds of action are possible, and how their fellow believers will react to mobilization efforts. REMOs must exercise great care and creativity to show how a given tradition's teachings about justice or charity should apply to ecological issues. About 20 percent of all emergence stories emphasized the critical role the tradition or set of traditions played. Many of these accounts began with some kind of conversion, awakening, or consciousness-raising experience and identified how a specific tradition helped individuals interpret events or experiences as a sacred call to get involved with faith-based environmentalism or create a new REMO.

During the early 1990s, Peter Ilyn, an evangelical minster, spent several summers hiking the Cascade Mountains in Washington. One day he ran into a herd of elk and two days later saw the largest clear cut he had ever seen—then he had an awakening. "I had this kind of juxtapositioning between the inherent goodness of the wild earth and the destructive power of the sense that there seemed to be no limits to what humans felt they had the right to do." As he notes on the Restoring Eden web page, "I went into the mountains a minister, but I came out an environmental activist."[22] This was also the period of the spotted owl controversy in the Northwest, which pitted the timber industry and wise-use movement against environmentalists seeking to protect the owl and its old growth Douglas fir forest habitat. After his wilderness epiphany, Reverend Ilyn decided to join the fight to save the owl and the forests, reasoning that "how can we be pro-life and yet cavalier with the very systems that create life? If we love the Creator, we must take care of creation." He searched for the biblical justification of his nascent environmentalism and found it in the Psalms ("In wisdom you made them all, the earth is filled with all of your creatures"—Psalms 104:24—"All that you have made praises you and your servants extol you—Psalms 145:10), and then began to develop an alternative evangelical environmental theology and plan of action.[23] He believed that there were many evangelicals who did not adhere to "dominion theology" (that God placed humans in charge of the natural world and allows them to use all plant and animal life for the benefit of humankind) or in the political program of the Moral Majority, but that they remained silent or unengaged because it was not safe for evangelicals to "love nature."[24]

He began preaching about Christians' duty to be good stewards to all of creation and sought out like-minded evangelicals with whom he joined lobbying Congress to preserve the Endangered Species Act. Out of this experience he formed Christians for Environmental Stewardship, which in 2001 became Restoring Eden. Both organizations sought to create a community for evangelicals that would help them reimagine their understanding of the natural world and God's call for them to be good stewards of it.

In his alternative view, nature is not a "well-oiled machine" that humans use for their own purposes, but "a system of living communities that scripture says are singing praise to the Creator, you know, be it the trees, the worms in the earth, the wild jackals" and a place that reveals the presence and grace of God. Restoring Eden was established to help evangelicals place themselves in an alternative biblical story about creation and then "tend and keep" it as they are called to be by their creator:

> All God wants is for us to love what God loves. . . . So it becomes, in a way, the work of Restoring Eden, and to help people recognize that their world view is not necessarily rooted in reality, it's rooted in story. And once you recognize story, you can either step back from it or you can rediscover your own story, or we can help you rediscover your story. . . . At the end of the day, a lot of what Restoring Eden tries to do is make it safe, theologically and culturally, for Christians to love nature and have that love of nature reinforce their faith, and in return, have their faith reinforce their love of nature.

This prototypical account begins with a wilderness experience in the context of a contentious environmental-industry fight over forests and endangered species, but filtered through Peter Ilyn's evangelicalism. Reverend Ilyn mined his tradition's theological warrants and hermeneutical approach (for example, proof-texting to find support for his environmental agenda in scripture) to stake out a critical and somewhat radical approach to faith-based environmentalism.[25] His embeddedness in comparatively more progressive evangelical circles played a secondary role as he sought to position Restoring Eden as an alternative to dominion theology or the antienvironmentalism espoused by conservative Christian Right leaders.

Mobilizing Networks

Scholars have studied the critical role interpersonal and interorganizational networks play in the activities of social movements, especially for mobilizing participation and garnering resources. Emergent movement organizations often tap into existing interpersonal and interorganizational networks because they trust members of the network, and because the networks serve as a source of information about opportunities, threats, and the framing of social problems.[26] Nearly 30 percent of the emergence accounts stressed the central role various networks played in the creation of their REMO. Many of these accounts discussed how founders activated the organizational networks to which they belonged, especially local associations of clergy, congregations within the same denomination or area, or other religious nonprofits. In some cases an existing network provided opportunities to organize or provided the funding necessary for a REMO to get established, while in others, a founder activated members of her personal networks who then brought their organizations into conversations about establishing a religious environmental group. In other words, activists' embeddedness in organizational and individual networks played an important role in the emergence of the new movement.

Paul Gorman had long been involved in progressive politics as a staff member for senator Eugene McCarthy and as an active participant in the civil rights and antiwar movements. In 1983, he was hired as the vice president of programs at the Episcopal Cathedral of Saint John the Divine in New York City. The church was a leader in social justice causes and interfaith activities in the city and often hosted such speakers as Jimmy Carter and Václav Havel. In the mid-1980s, a variety of thinkers began developing environmental theology, or what Gorman calls "sacred ecology." This new theology captured the imagination of the dean of the cathedral, and the church soon began offering programs on the environment. In 1986, Gorman was invited to attend the twenty-fifth anniversary of the World Wildlife Fund that focused on religion. There he interacted with a "who's who" of international religion and science, and a few years later he would activate those relationships. But first Gorman had his own awakening experience that he claimed was a call to start a religious environmental movement. The cathedral organized a special ritual to celebrate the feast day of Saint Francis of Assisi in which animals are brought into the church to be blessed, and among the animals Gorman brought was an elephant. He described the event as a watershed movement that pushed him toward engaging more deeply in religious environmentalism:

> We wondered what would happen when we opened the two-ton bronze doors of the Cathedral, which are spaces you could fit the Statue of Liberty in, you know, underneath the dome. Here's the image—so the two-ton bronze doors, which only open at Easter and it takes like four people to open, come open and there's this silhouetted elephant. What would be the reaction when that procession comes out of the nave, you know, flashbulbs or cheers or stampede or the things people do. And when it happened, and every year afterwards, the place went profoundly silent, and people, including me, began to choke up and tear. And the question went through my head—what are these tears? And it's a wonderful experience. Is it a reconciliation of the nature and faith? Who locked the doors in the first place? Who's blessing whom? How was it that the sacred is all sort of inside, and the profane is outside, and all of that. In a certain sense, everything we're talking about here is about that almost biblical moment of not quite restoring Eden, but re-imaging it. So it just became kind of clear to me having been involved in the social movements of the '60s

and '70s, that we needed to move this vision of sacred ecology into a religious community that has very little engagement with environment.

Just ten months earlier, thirty-two Nobel laureates and other eminent scientists had issued "An Open Letter to the American Religious Community" in which they called on all faith traditions and religious bodies to join in the fight to stave off the growing environmental crisis brought on by global warming. Gorman, now energized by the Saint Francis ritual, tapped into his networks to organize a meeting between scientists and religious leaders in Moscow (including Carl Sagan and Senator Al Gore). After the Moscow meeting, Gorman learned that few national religious bodies in the United States were working on the environment. No one seemed to staff or provide resources to take on climate change or other major ecological problems. Moreover, the theological foundations for religious environmentalism were in their infancy, and Gorman recognized that much more work needed to be done. He noted:

> We had to get the theology right so it's not New Age, that this is orthodox stuff for us all. We all know that this just isn't in the Book, it's in the original Book [i.e., Genesis]; it's like in the first chapter and that's striking. And part of it is to get the theology right because you are not even going to get in the door, past gatekeepers or skeptics, if you don't do that.

In order to "get the theology right," Gorman drew on his contacts within government (for example, Senator Al Gore), the natural sciences (for example, Carl Sagan), and a spectrum of American Judeo-Christian organizations, and invited them to a Washington, DC, conference in 1992. He carefully orchestrated the event so that Jewish, Catholic, and Protestant (mainline and liberal traditions, Evangelical and historic black traditions) participants were allowed to discuss their own tradition's perspective on environmentalism without the need to censor themselves. This safe space allowed participants to identify common ground and develop unified positions they could bring with them. Gorman also brought delegates together for interfaith discussions and events. One night Carl Sagan and Al Gore arranged to show the IMAX movie *Blue Planet* to all of the delegates as a way of provoking conversation and galvanizing support for Gorman's hope for greater religious involvement at the national level. The next morning at a meeting in the National Cathedral, the National Religious Partnership for the Environment moved closer to fruition:

The presiding bishop of the Episcopal Church, the first guy, gets up and says, "Until I saw that film last night and prayed about it afterwards, I didn't realize the depth of my love of creation, and my deep distress and its current condition," and then he sat down. And that was another one of those opening-the-door moments that I never, you know, it was chilling. And I remember thinking "this is irresistible." There was a whole day meeting, and we basically identified some of the priorities or values that would be important for the religious community to explore, albeit in our different groups. This is the advantage of everybody largely starting out from scratch together, which is not usually the case in religious lives. . . . This is new, but you know it is inescapably religious. It affects the religious imagination.

Following the meetings, Gorman and Evangelical, Protestant, Catholic, and Jewish leaders settled on a structure and plan of action for the new organization. Gorman was charged with serving as the NRPE's director and would help each of the four groups devise their goals and activities; raise funds; liaise with governmental actors, foundations, and the press; and coordinate advocacy work.[27]

Gorman's account highlights the important role social networks played in the emergence of some REMOs. He used his own networks and tapped into those of his colleagues to pull religious actors into dialogue and collaboration. The genius of the NRPE was that its founding members did not require any one group to compromise on its theological commitments and essential religious identities. Instead, the consultations connected religionists with others from their own tradition; collectively they then studied their own traditions and determined how best to act on their tradition. The story of the NRPE, like that of Restoring Eden, also highlights how biographical predispositions (in this case Gorman's move from politics to religion) are critical for the emergence of REMOs, especially in the way that immersion in what Gorman calls the "religious imagination" or in networks that offer new ways of thinking about the sacred and the environment, pushed the founders to take their initial steps toward organizing the movement. Finally, both stories highlight how REMO founders drew on the narrative of conversion or awakening that play an important role in Christianity and suggest that religious activists may be embedded in both intensive and extensive cultural schemas.[28]

"Piecing Things Together": Merging Traditions and Networks

In the most common emergence story (about one-third of all) both embeddedness in a tradition and within a particular religious network are equally important. Many of these REMOs emerged as individuals acted within local congregational, clergy, or religious nonprofit networks and aimed to harness their particular tradition to galvanize interest and support or blend traditions to appeal to members of different communities. In some cases the tradition played a sensitizing role as it helped believers understand why they had a religious obligation toward the environment and a motivating role as they sought to persuade others within their community of this newfound spiritual calling. Thus religious traditions became important tools for building and/or mobilizing a given religious network. This type of emergence is illustrated in the following account of the Partnership for Earth Spirituality (PES).

When I asked Sister Joan Brown, founder of the PES, how she got involved with environmental ministry, she immediately but gently corrected me:

> I call it "ecology ministry." This is the reason—environment with ordinary people is a red flag. And ecology speaks more of the spiritual and ethical nature because I love the root of the word ecology, which comes from *oicos*, which is also the same word as economy and it means "home." But basically, I'm making relationships and caring for relationships and caring for this home that we've been given, which is a sacred task. So I call it ecology ministry.

She continued by briefly describing her thirty years of ministry as a member of the Order of Saint Francis working on a host of issues related to peace, nonviolence, and economic justice. She noted in the interview how she "began to see more clearly that non-violence in humans needs to be connected with creation, and then moving liberation theology from a preferential option for the poor to the preferential option for the economically poor and for the earth." In short her foundation in Catholic social ethics and liberation theology, and years of working for justice, led her to focus on the issue of sustainability. As she worked with two of her sisters on sustainable housing along the Mexico–New Mexico border, she came to see that there were "a lot of structural obstacles within society for moving in a practical way towards living sustainably. I saw that there was a deeper thing we needed to change."

Brown's religious order provided her with significant autonomy to pursue her interests in environmental ministries and growing support within the Catholic hierarchy for environmental justice.[29] First she sought additional education in eco-spirituality and ecology at the California Institute of Integral Studies and with Thomas Berry (the leading Catholic eco-theologian).

Upon returning to her community in New Mexico, she asked the Social Justice Office of the Catholic Archdiocese of New Mexico for help developing a diocesan program on sustainable housing. During her work on sustainable housing, she came to realize that many of the obstacles toward living sustainably were grounded in the "anthropocentric" and dualistic worldview of Americans. She decided that she wanted to start a new organization to provide a more holistic and eco-centered way of thinking about the religious life and the natural world, and to develop religious rituals that would more firmly connect humans with the natural world and engender a deep passion for people of all faiths to care for creation. She hoped that by teaching and promoting a contemplative life that was focused on the Creator and the natural world, religious individuals and groups would be motivated to take action to preserve the environment. She did not imagine at the time that her new organization would become a center of activism and protest in the state.

New Mexico is home to a vibrant Catholic community but also to indigenous religious communities and a host of individual spiritual seekers, many of whom, in the 1980s and 1990s, were keenly interested in environmental sustainability. Thus Brown saw an opportunity to bring various groups together under the umbrella of an interfaith organization. With the support of her community and the archdiocese, Sister Brown initiated a series of conversations with Judeo-Christian and Native American leaders across the state to identify the range and depth of support for the proposed partnership. She and a number of "partners" developed a vision of earth spirituality and determined that there was indeed support to create a center for eco-spirituality. The proposed center would provide educational and ritual resources and engage in advocacy around environmental issues facing the state. Originally she hoped it would be housed at her religious order, but financial constraints within the order led her to cancel this plan. Undeterred, she and her partners decided to create a "virtual organization" in which she and other board members would organize eco-spirituality retreats, offer courses at congregations, and develop an advocacy network that they could activate via listservs and phone trees. PES was soon involved with a variety of nonreligious environmental and Native American groups around issues of climate change and energy, including protests to stop the opening of a new uranium mine and the construction of a power plant on tribal lands. PES also joined the New Mexico Coalition on Climate Change and supported the Southwest Organizing Project's (SWOP) lawsuit against the Department of Energy for violations of the Clean Water Act at Los Alamos National Laboratory. Many of their political actions were bookended by religious rituals that often blended ideas and elements from Catholic, Native American, and eco-spirituality traditions. Rather than adopt the "walk together separately" approach of the NRPE, PES combined various traditions to build consensus and support around its mission. Sister Brown was able to blend multiple traditions and create a new green ethic because her partners shared theologically similar perspectives (unlike the members of the NRPE) and because she was deeply connected to a variety of religious traditions (Catholicism and various strains of eco-spirituality). In some ways, her religious networks exposed her to these various green traditions and encouraged her to blend them together in the work of the new REMO. In sum, cultural and relational embeddedness worked symbiotically to help the PES get established.

Expanding the Mission

The fourth type of REMO emergence corresponds to the groups of religious nonprofits that were already in existence but did not originally include the environment in their activities. About one-fifth of all REMOs fall into this category. Most of these organizations were involved in faith-based advocacy and service provision (such as the Maine Council of Churches, Ecumenical Ministries of Oregon, the National Catholic Rural Life Conference, and the National Council of Churches) but did not add the environment to their portfolios until a staff or board member persuaded the organizations' leaders that it was an important issue in which to get involved. In most cases their environmental justice ministries have become a central part of the organization's work, often with dedicated staff and financial resources. The emergence story of the REMOs is characterized by the expansion of an organization's mission as key staff or board members engage in an internal organizational process of aligning or extending the existing mission to include ecological issues. Adding the environment seemed like a "natural" next step for many as it was consistent with their extant concern and work for the poor, for example, and it dovetailed with the interests of their members and supporters. The addition of environmental activism by Progressive Christians Uniting is a good example of this type of emergence and illustrates how place-specific ecological issues, a particular tradition, and organizational networks all played a role.

As its name implies, Progressive Christians Uniting (PCU) is a liberal Protestant organization. Founded in 1996 in southern California by an activist Episcopal priest (Rev. George Regis) and theologian John Cobb Jr., the PCU was established as a progressive Christian response to the Christian Right. It worked on a variety of justice issues including the morality of capitalism, labor rights, affordable housing, immigration, and GLBQT rights. Since the early 1970s, Cobb has written on the environment, and he is considered one of the originators of eco-theology, but the PCU did not tackle the environment in a programmatic way until 2006. Jennifer Snow, a staffer who worked primarily on the administrative side of the organization, learned about the impending closure of a large community garden in Los Angeles and received permission to get involved as a representative of PCU. In her account of how the struggle to keep the garden open served as the catalyst for the PCU's Eighth Day Project, Snow echoes the justice concerns that motivate the organization and, somewhat surprisingly to her, the absence of religious groups in the fight:

The south central farm was the largest urban farm in the United States. It was 14 acres. After the LA riots, the city gave the land to the LA food bank for a community garden. So it was this huge, successful community garden—360 families had plots there. A lot of them were Latino immigrants, and they had all kinds of heirloom plants and seeds they had brought with them from Central America, and they grew medicinal plants there. It was a place where they could get healthy food, which otherwise is not easy to come by in that neighborhood. It's all liquor stores and convenience stores. It was the only green space for blocks and blocks. It was what they call the Alameda Corridor, which is where the trains from Long Beach, from the port, and all of the diesel trucks from the port, they drive right through this neighborhood. It's so polluted, and there's no green space at all, except for the farm. . . . But in 2003, the city sold the land back to the original developer for $5 million, which is a fraction of what it's worth. The developer immediately said he was evicting the farmers. So for the next year, year-and-a-half, there was a 24-hour vigil at the farm. . . . And what was amazing was that there were people from all economic backgrounds, all races, all cultures, all ages. There was a deep sense of spirituality and spiritual commitment and joy. *What I noticed was that the churches were not there* [emphasis mine], and that struck me because, you know, I'm a Christian; I know some of my friends who are Christian, but for the most part, the activists felt as though the Christian community was not their friend. And the Christian churches were not visibly there at all, like no church sent representatives that said, "Yes, our congregation is behind you or our denomination is behind you." I thought that was really striking because I work with progressive churches, and I know that they're all environmentalists, but yet when they're faced with this really serious environmental issue, right in their own backyard, and furthermore one that's really connected to issues of economic justice, it just went right over their heads.

In the end the protesters were evicted and the community garden was shut down. Incensed and saddened by this event, Snow decided that churches in the region needed to become aware of and active on issues that profoundly affect the poor and the earth. She wanted to "create a network" of churches and individual Christians that would "take up the responsibility of caring for creation, sharing resources justly and peaceably with our neighbors."[30] Snow created this new program in extra time she had to carve out of her already full agenda at PCU. The name of the program, the Eighth Day Project, alluded to the Genesis account of creation in which humans are expected to start caring for creation after God finished God's week of creating the universe. It also underlined the liberal Protestant commitment to justice and practical ethical action. This commitment is evinced in her summary of the program's mission: "The goal of the Eighth Day Project is to educate and nurture this commitment to faith-based activism among Christians, and our hope is to see progressive Christians and their congregations living and struggling with this commitment as instinctively and as naturally as they work for peace and economic justice in other areas."

In short, PCU delved into environmental issues because one of its staff members, who was embedded in a particular eco-space and in the liberal Protestant tradition, was sensitized to an environmental problem in Los Angeles. She also worked for a religious nonprofit already engaged in a variety of justice issues and whose leaders were open to adding the environment to its agenda. She then harnessed the resources of the organization to expand its mission in an attempt to prevent future environmental injustices of this kind. The new mission and the project were set-up to be aligned with the larger set of religious values and activities of the PCU and its constituency. The tradition helped Snow recognize the injustice surrounding the closing of the community garden and the appalling quiescence and absence of the Christian community, and prompted her to develop a rationale and frame for the new program that would effectively mobilize progressive churches and individual Christians in the region.

Conclusion

In this chapter I have developed an alternative explanation of movement emergence—one that moves beyond the canonical accounts from the political opportunity and resource dependency theories. I showed that the former, in particular, is not particularly salient, as faith-based environmentalism did not emerge because of new opportunities in the political sphere. Politics, especially at the national or macro levels, rarely entered into founders' calculus, and the decades-long process of emergence suggests that founders were not trying to capitalize on new vulnerabilities within the political system or gain the help of political allies. Rather, this new form of environmental activism emerged as a distinctly religious phenomenon following a distinctly religious dynamic. While large-scale environmental and political conditions (for example, global warming) were important, this chapter has shown how founders' embeddedness in specific religious traditions led them to interpret the larger environmental situation, place-specific ecological problems, or even their own experiences of nature in ways that motivated them to establish faith-based environmental organizations. Thus religious processes of conversion or awakening, rather than political opportunities directly, initiated the creation of REMOs, and as I discuss more fully in the next chapter, REMOs emerged in part because deeply religious individuals saw environmental action and advocacy as a way to be more authentically religious and to help their particular communities more authentically live out their religious mandates. Founders' embeddedness in their given tradition—their commitment to ethical principles about justice, stewardship, and creation, and concomitant religious practices—led them to take environmental action because it was their religious duty. Moreover, their grounding in particular traditions provided them with the cultural or spiritual resources they would use to mobilize support from their religious communities. Although REMOs emerged with far fewer financial and material resources than other movements, the cultural resources on which they drew provided "models of and models for" creating a new movement.[31]

I also showed how interpersonal and organizational ties were crucial in the emergence of the new movement. Some REMOs emerged from a process of discernment after members of an existing religious network had participated in a campaign of some secular environmental movement organization, while others were established because members of their network refused to act. REMO networks provided access to resources (albeit limited) and opportunities to figure out why they should get involved and how to organize. The new REMOs mimicked one of two existing forms widely followed by religious groups: coalitions of theologically or religiously similar groups and individuals (like local clergy roundtables or the Alinksy type of local faith-based organizations described by Warren, Wood, and Ammerman) or coalitions of different groups that may share a common concern with some ecological problem but that are formally free to act collectively or independently as dictated by their tradition and constituency.[32]

In the next chapter, I deepen this embeddedness perspective by turning to an examination of how REMOs developed and gained a foothold within the American religious sphere. I focus on the strategic and entrepreneurial choices the organizations' leaders made as they created their unique missions, repertoires of action, and ultimately a new form of niche activism. In particular I show how organizations' embeddedness in specific religious traditions, networks, and audiences (both real and imagined) constrained and directed this process.

THREE

Mission, Strategy, and the Search for Legitimacy

Like David facing Goliath, Jonathan Merritt faced a set of seemingly insurmountable challenges as he looked to start the Southern Baptist Climate Initiative (SBCI). First, he needed to persuade members of his conservative Christian denomination that climate change was real; second, that environmentalism was a legitimate religious issue and not one simply the bailiwick of liberal pagan activists; third, that the Bible spoke about nature and ecological issues; and finally, that their faith called on them to take action. Mobilizing conservative Protestants has been problematic given the long-held suspicion within this community that the environmental movement is generally irreligious and, more specifically, anti-Christian. Yet Merritt's task was made more difficult by the resistance he met from the leadership of the Southern Baptist Convention (SBC). As noted in chapter 2, he initially worked with five respected SBC seminary leaders in order to "get the language right," because he expected that once he went public with the initiative his fellow Baptists would "just tear things apart." Merritt hoped to gain the support of a small number of influential SBC leaders (seminary and college presidents, megachurch pastors, and SBC agency heads) who would then urge congregations and individual Baptists to make climate change a major social and ethical issue of the denomination. When dozens of leaders signed the statement in support, the SBC's Ethics and Religious Liberty Commission (ERLC) pressed signers to withdraw their support.[1] When I asked him to explain why his organizing efforts were resisted by the ERLC and, more generally, by members of the denomination, Merritt discussed how the root cause was located in the close alignment of the environment with liberal politics:

> The environmental issue is the tip of the iceberg [for many conservatives].
> It's sort of an entry point for people to become engaged in a broader moral

agenda. . . . For example, if you are an environmentalist, you have to be concerned with malaria, and you have to be concerned with poverty and hunger. And these are traditionally more liberal political issues. So they feel that if you open up this door, and they speak in no uncertain terms about these things, if you open up this door, then you're opening up the door to an entire agenda that runs counter to their political priorities. In fact, when I was on ABC *World News* shortly after the Declaration, they interviewed Tony Perkins of the Family Research Council, and Tony said, "Well, if we start, opening up the door to climate change and environmentalism, these people will want to have population control; then that means forced abortions, and encouraging homosexual marriage." And I'm thinking "Wow, whoa, what a non-sequitur; we're not even there yet." So I really do believe that the connection is inevitable between more liberal issues and this issue, which is why they fight so vehemently for it.

According to Merritt, Baptist ethics and politics are embedded in the larger "culture wars," and thus any issue championed by the left is seen as illegitimate. He also noted in our interview that for many leaders of his church, theology and political ideology are closely aligned; to be conservative theologically must also mean that one is politically conservative or Republican. The political differences between Merritt and the leaders of his denomination also map onto important generational differences. This divide runs between older Baptists, who emphasize orthodoxy or right belief, and younger (under forty) Baptists, who emphasize orthopraxy or right action.[2] In our interview he talked about how many younger evangelicals self-identify as "red-letter" Christians whose faith starts with the ethical teachings and example of Jesus rather than with a "particular party line or theological paradigm." The term "red-letter" refers to a version of the Bible in which the words of Jesus are printed in red. By adopting this moniker, younger Baptists demonstrate their commitment to a literalistic hermeneutic and way of being Christian that prioritizes action or ethics. This kind of biblicism is a common part of conservative Protestantism and anchors the SBCI in its historic tradition. The formal declaration the SBCI wrote is peppered with citations to Southern Baptist doctrine—largely from the 2000 document *Baptist Faith and Message*—as well as to biblical warrants that clearly ground the call to environmentalism in the tradition. The Declaration also serves to depoliticize Merritt's work to raise awareness and mobilize Baptists to take action:

We must care about environmental issues because of our commitment to God's Holy and inerrant Word, which is "the supreme standard by which all human conduct, creeds, and religious opinions should be tried" [*Baptist Faith and Message*, 2000]. Within these scriptures, we are reminded that when God made mankind, He commissioned us to exercise stewardship over the earth and its creatures (Gen. 1:26–28). Therefore, our motivation for facing failures to exercise proper stewardship is not primarily political, social, or economic—it is primarily biblical (emphasis original).[3]

According to Merritt, the emphasis on ethics driven by a literal reading of the Gospels, made popular in the 1990s by the "What Would Jesus Do?" movement in evangelical youth circles, has helped push young Baptists to take up the environment and global warming as important religious causes. Even though older Baptists may agree that humans have a religious duty to be good stewards of creation, this does not necessarily lead to agreement on the kinds of action Baptists should take:

> I'm reading a book that's out of print right now by Richard Lamb[4] called *The Earth Is the Lord's*. It's about what Southern Baptist leaders wrote about environmentalism in 1992. They all affirmed the same things I affirm. If you say, "Yeah, the earth was created; yes, he called it very good; yes, he's made us stewards of it; yes, it's God's revelation—check, check, check, check, check." Okay, what do we do about it? Well, now we're going to diverge. I'm going to take a more aggressive stance than they're going to take because their orthodoxy leads to a totally different practice. The way that their orthodoxy plays out in real life is totally different than with me.

The generational challenge Merritt faced then was, first, to give voice to the environmental and theological commitments of younger Baptists; second, to have that voice recognized and granted legitimacy by the denomination's leaders and older members; and third, to ask the denomination as a whole to change its positions and actions about global warming. Merritt created the SBCI to mobilize networks of influential SBC leaders and young Baptists. His hope was to start an internal conversation among Southern Baptists, to prod them to "think biblically" and "to bridge the gap between orthodoxy and orthopraxy." If successful, the SBCI will first educate and raise the awareness of Baptist elites, who in turn will help Baptists of all ages "make the leap from right thought to right practice." Merritt's account illustrates some of the challenges REMOs who work within a single tradition, especially conservative Protestant traditions, may face when they attempt to gain legitimacy and mobilize support. The next story highlights how interfaith REMOs work to blend or speak to diverse traditions, and suggests how

REMOs, more generally, seek to carve out specific niches within the larger movement field.

GreenFaith (GF), in New Jersey, was founded as an interfaith organization by Jewish and Episcopal leaders who were convinced that the religious communities in the state needed to start discussing how faith and nature were related. This conviction grew not only from their experiences at the 1992 United Nations Conference on Environment and Development (or the Rio Earth Summit), but also from witnessing the wide range of environmental problems in the state—from overpopulation and suburban sprawl to air pollution and dozens of toxic waste sites that pose significant health risks for poorer populations. According to GreenFaith's website, "New Jerseyans— including children—in two-thirds of its counties breathe air that does not meet federal health standards. Pediatric asthma is widespread. Every New Jersey resident lives within ten miles of a Superfund site."[5] GreenFaith's leaders also recognized that the state was religiously diverse and began to look for common values and beliefs that could unite them into a powerful force to restore the state's ecosystem.

Eventually the REMO identified three core values shared by the world's religions and built its mission and programs around them: stewardship ("using natural resources carefully and minimizing human impacts on the environment"), justice ("all people have a right to a clean and healthy environment"), and spirit ("we believe that religious traditions see the sacred in nature and that people grow spiritually through a strong relationship with the earth").[6] Rev. Fletcher Harper, GreenFaith's executive director, explained that "these values are very widely acceptable to a significant range of different religious traditions" and therefore make the organization salient to a wide religious audience. The REMO's mission is to use these broad values to "inspire, educate, and mobilize people of diverse religious backgrounds for environmental leadership."[7] GF focuses much of its work on education. They offer numerous courses that aim to help religious individuals learn about what their traditions teach about the environment and about specific ecological issues. They have regularly conducted environmental justice tours through some of the most polluted areas of New Jersey to help lay and ordained congregational leaders see firsthand how pollution impacts vulnerable populations, and they offer workshops on divesting in fossil fuels and greening religious buildings and provide congregations with tools that will help them learn how to engage in advocacy for ecologically at-risk communities.

Leadership has become increasingly central in its work as the REMO is trying to create a national cadre of religious leaders who will promote and work for religious environmentalism in their own eco-spaces. In 2007, GF developed the first national fellowship program to train lay and ordained individuals "for religiously based environmental leadership." The rationale and hope for this program highlight the ambitious plan to reshape American society, much like the civil rights movement did:

> At GreenFaith, two beliefs most powerfully drive our work. First, *we believe that the vast majority of people experience the sacred or sublime in nature, and that the depth of such experiences creates common ground for people of diverse traditions in relation to the one earth we all share.* Second, *we believe religious communities can mobilize large numbers of people to take action for the earth* [emphasis original]. These communities have mobilized millions in relation to other issues in the recent past—the Civil Rights Movement comes immediately to mind. We see no reason to believe this cannot happen in relation to the earth. So we offer the *GreenFaith Fellowship Program* with the very real hope that our work on this Fellowship can create a brighter future for the earth and for religion.[8]

Despite the success of GreenFaith's various programs, it has not been easy. Harper discussed the reticence of religious groups to get involved (as opposed to the kind of active resistance Jonathan Merritt described):

> Religious leaders are not well educated either scientifically or environmentally, which means that they feel unable to make sound judgments about what types of issues ought to be a priority. I think it's a real bind that a lot of religious groups are in; they just don't have the staff people who understand how to identify high-level priorities environmentally. So I think that's a major issue. I think the second issue is created by a historic lack of connection. I mean religious groups have not been involved in environmental issues in any significant way; whereas, religious groups have always been involved in issues around poverty or hunger. So it's just that environmental stuff is very new to religious groups, and they just don't generally feel comfortable about it. I think that another piece of the picture has been that, and this is especially true in the Christian community—there's an often unspoken assumption, and in Evangelical communities it quite frequently is spoken, that environmentalists sort of have their own quasi-religious system set up and that it tends to be pagan and tends to be not terribly receptive to Jewish or Christian traditional expressions of Christianity or Judaism.

Harper noted that this assumption was held by many within mainline Protestantism and has prevented these communities from making the environment a pressing religious concern. More generally he argued that the novelty of the environment as a pressing religious concern and the lack of theological and practical resources on which to draw have held religious leaders and their congregations back.

These two accounts highlight the challenges REMOs face as they attempt to develop and institutionalize religious environmentalism. Rao and his colleagues argue that the creation of a new organizational form, such as a REMO, is a political venture insofar as the founders must "actively define, justify, and push the theory and values underpinning the new form" in order to garner support from existing players in the institutional field, wrestle resources away from them, and gain legitimacy for the new endeavor.[9] In other words, the leaders of the nascent movement organizations faced the challenging task of creating statements about their missions, goals, and identity that would persuade a skeptical, indifferent, and even hostile audience that this new endeavor was religiously legitimate. And it was an audience that often possessed considerable power it could use to help or hinder the new movement organizations (for example, nationally recognized megachurch pastors).

Concerns about legitimacy were expressed in most of the interviews, as evinced in Paul Gorman's comment about "getting the theology right" in order to get past denominational gatekeepers, and in Pat Hudson's discussion of LEAF's conservative Appalachian audience. Scholars who write about the emergence of new organizational forms consistently note that "institutional entrepreneurs" (the founders of new organizations) must establish the legitimacy of their innovations if the new organization is to survive.[10] New organizations must be seen as conforming to or at least being congruent with the operative rules, practices, and cultures of organizations within the extant field in order for their founders successfully to accomplish the other tasks involved in creating new organizations, such as gathering resources, creating a shared identity, and setting organizational boundaries.[11] McGuire, Hardy, and Lawrence contend that institutional entrepreneurs' success lies in their ability to "connect their change project to the activities and interests of other actors in the field, crafting their project to fit the conditions of the field itself."[12] Similarly, Ruef argues, "Novel organizational forms are most likely to become legitimated when they fit into the preexisting cultural beliefs, meanings, and typifications of an organizational community."[13] In short, gaining legitimacy depends on how well entrepreneurs are able to connect old ideas, practices, values, or institutional logics with new ones, and they do so by crafting accounts that explain and justify the new organizational form or set of practices.[14] These accounts must resonate with potential supporters and key stakeholders or actors who can provide critical support for the new organization (denominations or congregations in the case of REMOs).

The task of gaining legitimacy for the nascent religious environmental movement was especially challenging for several reasons. First, environmentalism and direct political action have been considered illegitimate or at least not very important to many religious bodies and individuals. Environmentalism has been so closely aligned with secular liberal politics that it was difficult to even raise it as a serious religious issue in many communities. Second, much of American Christianity was immersed in battles over cultural issues (that is, abortion and homosexuality) and/or busy with a host of human-focused benevolent activities (for example, charity), and thus had little interest in or energy for adding another "cause" to the long list of issues on which congregations and denominations worked. Finally, the newly created REMOs were attempting to create a new organizational space in between the secular environmental movement and the field of American religion that is anchored in denominations and congregations. Each institutional space follows a distinct logic or a set of rules that sets out the mission and goals; defines problems and possible solutions; and identifies the legitimate organizational forms, strategies of action, and constituencies for actors within a given field.[15] REMOS fall between the fields of American religion and American environmentalism, which creates unique problems for gaining legitimacy and establishing the new movement: "The project of creating a field-wide environment for a new form becomes more challenging at the intersection of multiple organizational fields, due to diverse interests, multiple (often competing) frames, and entrenched sources of resources in established fields."[16] If REMOs follow the logic of the secular environmental movement too closely, they risk alienating religionists who represent their core constituency and source of resources. Conversely, adhering to a religious logic too tightly might alienate secular environmental groups that could be an important collaborator and source of help since many REMOs lack the requisite scientific knowledge, political acumen, or financial resources to become successful players in the larger social movement field.

Organizational Missions

A mission statement is an important tool for new organizations because it sends a message to external constituencies and potential supporters about shared values and goals, identifies the organization's boundaries, and sets forth its distinctive contributions to larger efforts to affect social change.[17] Statements about organizational mission and goals are not simply exercises REMOs follow in order to conform to some normative requirement for religious nonprofits, as neoinstitutional explanations suggest. They are critical means for REMOs to present themselves as legitimate in the eyes of their real and imagined religious audiences. REMO mission statements provide answers to questions that establish their identity and legitimacy: What kind of organization are we? Why will we engage in education or consciousness-raising but not protests? What are the connections between religion and nature, and does our particular religion command us to act in environmentally sound ways? Why must Baptist, Jews, or Buddhists engage in environmental action in order to be fully and authentically religious?

Although they are thoughtful and creative, REMO leaders do not develop these statements or construct their new organizations however they wish. They are constrained by the ways in which they are embedded in particular religious traditions and networks. This is illustrated in the pains Jonathan Merritt took to show how environmentalism was mandated by scripture and followed from a particular kind of evangelical interpretive strategy commonly used among his audience. It is evident, as well, in the decision by GreenFaith to establish an interfaith organization that could mirror the religious diversity of New Jersey. Perhaps more importantly, REMOs' embeddedness in specific religious traditions and networks or religious fields means that they must follow or at least acknowledge the preexisting set of rules of the game and ways of seeing the world that govern the organizations in the field. Changing these rules and practices may be difficult because they are institutionally entrenched and have a taken-for-granted character.[18] For example, most religions are engaged in activities to alleviate human suffering, and this commitment may make it difficult for them to recognize environmental problems as salient to their religious missions. Similarly, the strong association between the environmental movement and liberal politics and the movement's alleged hostility toward religions may make many religious groups wary of getting involved, as both Merritt and Harper suggested. Thus the imperatives to demonstrate that REMOs share the same religious values as their audience, and at the same time distance themselves from secular environmentalism, guided their missional work.

Many REMOs are strongly embedded in specific denominations or traditions, and they capitalized on their religious ties or knowledge of a tradition to overcome institutional resistance to their endeavors. They did so by creatively mining and adapting elements from particular traditions to show how religious meaning systems justify environmentalism but did not alter the traditions so much that they became unrecognizable to their coreligion-ists.[19] Not surprisingly then, institutional, cultural, relational, and ecological contexts played an important role in the formation of the new movement organizations. In the rest of the chapter, I discuss how embeddedness in these different contexts led REMOs to develop their own missions as well as a type of niche activism. I first turn to an analysis of REMOs' mission state-ments and discussions of their organizational identities, and identify several features that cut across specific traditions or religious bodies and suggest how more general rules, ideas, and practices of the larger American religious field shaped the development of REMOs. Then I discuss how REMOs' em-beddedness in specific contexts led them to create or embed themselves in different movement niches.

Promoting a Religious Awakening

Religious environmental organizations focus their attention on mobilizing, educating, and supporting religious communities and their members. Their overall goal is to make religious communities aware that environmental problems are, at their root, religious problems and thus demand a religious response. REMOs are not trying to turn religious people into environmental-ists so much as to help them understand how being ecologically aware and active is fundamental to being faithful within a given tradition. For example, the Interfaith Network for Earth Concerns (INEC) identifies its mission as awakening people of faith to God's call to care for creation: "Our mission is to connect, inform and empower people, congregations and religious in-stitutions to work for justice and the care and renewal of the earth. Our aim is to foster an awareness that care for creation is integral to a life of faith."

This ideal of integrating an environmental ethic into the faith lives of in-dividuals and communities may be the most important goal of faith-based environmental organizations. Interviewees, especially those whose organi-zation's primary audience is congregations, frequently echoed the INEC's goal. They also spoke about the need to avoid the common problem of compartmentalizing the environment and making it the special purview of just another church committee or a small group within a congregation. Instead, their goal is to help religious communities make environmentalism integral to a life of faith. Rev. David Rhoads, the executive director of the Web of Creation, offers a clear summary of this goal:

> Our goal is to transform the life and mission of the whole congregation.
> Therefore, we have five areas that we work at—worship, education, building

and grounds, discipleship of the members individually (at home and work), and public ministry/political advocacy. We want each congregation to have a green team—the green team in each congregation is not really a separate committee that just does the environmental thing. Their goal is to be leaven—to get the worship committee to do environmental worship or care for creation worship; to get the buildings and grounds committee to be responsible for making some environmental changes.

Paul Gorman of the National Religious Partnership for the Environment (NRPE) claims that, like many other faith-based groups, the NRPE is attempting to awaken the moral imagination of individuals, congregations, and larger religious bodies to the current environmental crisis and to encourage religious groups to mine their own traditions for solutions to ecological problems. But more pointedly, he says, they want to "weave environmental vision and values across the entire fabric of religious life, once and for all." In particular the NRPE's programs and educational materials emphasize how the values of various traditions call on communities of faith to protect the totality of life and suggest how these green values can be given a more prominent role in the formation of religious identity. He summarizes the goal of awakening and revivifying the faithful in the document that explains the NRPE's climate change initiative, "God's Climate Embraces Us All":

> How can religious life help make climate change a moral issue? Deep religious values are having fresh power for people of faith in that first generation to behold the whole earth, precious and in peril. In Genesis, God designates creation as "very good" (Gen. 1:31) and commands us to "till and to tend the garden" (Gen. 2:15). In Psalms we read, "The Earth is the Lord's and the fullness thereof" (Ps. 24:1). We have a paramount obligation to care first for the "least of these" (Matthew 25:35) and to assure the future well-being of all life on Earth in God's "covenant which I make between you and every living creature for *perpetual generations.*" These values are flowing vividly into worship, sermons, and religious observance. They're inspirations, not frames, living articles of faith, not talking points. They are the reasons why care for God's creation is becoming the most compelling new cause in religious life and why global climate change is the issue that is most moving people of faith to act.

Here the use of Old Testament texts and the reference to God's "covenant" are important ways to communicate the NRPE's identity as an interfaith organization since they will both appeal to its primary Jewish and Christian constituents. Moreover, Gorman is trying to show how the NRPE's mission is fundamentally different from that of secular groups by omitting references to direct political action, advocacy, or litigation (the most common activities among the major environmental movement organizations in the United States), de-emphasizing conservation, and instead using theological rather than scientific language.

Non-Christian REMOs also identify their mission in terms of promoting a spiritual awakening and do so by grounding their appeals in particular traditions. For example, the Buddhist-inspired Green Sangha draws on the core Buddhist practice of meditation and several key concepts such as oneness and nonseparateness, or the interconnectedness of all life, to explain its mission. A statement written by the group's founder, Jonathan Gustin, describes the root problem of environmental degradation and Green Sangha's solution in distinctly Buddhist terms:

> *Self-realization is the root of true activism* [italics original]. At Green Sangha we believe that the most efficient path to environmental restoration is not just a series of partial "fixes" at the physical level of the biosphere. We believe that the root "fix" is at the level of consciousness. For example, one Green Sangha action aims at eliminating disposable plastics. The problem is not just plastics per se, but the divided consciousness that can create disposable plastics in the first place. Even if we were 100 percent successful in eradicating all plastic bags, that would leave countless other problems. And what causes these countless problems? The idea that we are each discrete entities with defined boundaries separate from the rest of life. As long as dualism dominates our cultural consciousness, problems will continue to be created. How do we wake ourselves and others up to the fundamental unity of life? Meditation. At Green Sangha we meditate together. . . . Through meditation we become aware of that which is already abiding as primordial consciousness. At Green Sangha we believe that when we rest as silence and stillness, then our actions can flow naturally. We call this "True Activism." "True," because it is action rooted in the truth of non-separation. In this sort of activism, the illusion of a separate activist disappears, and what is left is oneness moving to care for itself.[20]

As these examples indicate, REMOs work to connect their missions with particular theological or religious concepts and ideals, and they also attempt to show how their goal of promoting a green awakening fits into the social ethics of its religious audiences. REMOs are thus engaged in a project of redefining what it means to be a practicing religious person or an authentic congregation. The Reverend Katherine Jesch, executive director for the Unitarian Universalist Ministry for Earth (UUMFE), describes how they link environmental actions with the more traditional human-centered activities of congregations (for example, programs to feed or house the homeless) and argues that in order to address these problems fully, congregations must also integrate a concern for the environment into their collective lives:

> One thing that we're beginning to do now is we're drawing the links; we're connecting the dots with other justice issues. So working on an environmental issue is not in competition with working on mental health or health care, in general, or housing—affordable housing? And part of that is the way we're doing the green sanctuary environmental justice project. I tell congregations, don't invent a new project. What is it that you already do for justice? Almost every congregation is active in some way. So my home congregation in Arlington, VA, has a very active affordable housing program. So I say, "Make that affordable housing green. Don't invent a new project; revise the project you already have, where you already have the energy and commitment of the congregation."

This move to integrate environmental issues, identity, and action with more common concerns and practices of religious organizations is also a move to emphasize ethics and de-emphasize belief or orthodoxy. This shift, already introduced in the discussion of the SBCI, is not restricted to evangelicals; it cuts across a wide swath of western Christianity, and it is the foundation for much ecumenical and interfaith cooperation.[21] David Radcliff from the New Community Project (NCP) argues that being religious (and an environmentalist) is primarily about cultivating a way of being in the world that recognizes God's call to live out the ethical demands of Christianity. He is critical of contemporary religion in which "faith" is peddled as some kind of magical panacea for personal problems because these types of religions overshadow the model and teaching of Jesus. He made this argument after describing a typical workshop he runs in which he shares experiences of seeing the poverty engendered by the extractive industries in the Amazon, and how the Christian faith connects to these issues:

> This is what God expects from us, not the religious hocus-pocus, you know, what passes for religion in our churches and on TV. But this is kind of real life, and you know, Jesus is a great model for that in the sense that he incarnated these things. He just didn't talk about them or theologize about them or whatever; he was really living out his passions and what I think are God's passions. So I try to make that connection that being religious doesn't mean the things we've often been taught that it means—it can mean these other things too.

REMO leaders note that beliefs are still important, but they identify how specific beliefs are connected to the environment and how individuals should act on those beliefs to protect the environment. One common practice is for organizations to hold up biblical warrants for stewardship and then encourage religious groups to act on the call to tend "God's Garden" as evident in the following comment from Rev. John Rossing of Earthcare: "Our philosophy is that caring for creation is one of the premier ways for people to show their love and respect for the Creator; that we believe, as Genesis says, God put us here to tend and keep his garden." Pam Richart, co-executive director of the Eco-Justice Collaborative, argues that their work is firmly grounded in the theology and stories of traditional Protestantism, but that they change the valence or focus of the story in order to motivate action:

> We find there's a lot of value in connecting with people over stories and theol-
> ogy that they've grown up with, and twisting and turning it around to cause
> them to look at it differently. That's what we do. And it is the most awesome
> thing—to take something as simple as the story of the manna, . . . and turn it
> into a justice piece; one in which God is saying, "Take only what you need."
> God is saying, "If you take more than what you need; it's going to smell like
> maggots," right? And the redistribution part comes along in that story to care
> for the earth and also trusting that you're going to have enough, rather than
> accumulation. . . . And people go "aha."

While the study organizations share this broad mission to awaken communities of faith, they tailor their specific mission and goals to the particular constituencies they are attempting to mobilize. A handful of groups serve a specific denomination—Presbyterians for Restoring Creation, Unitarian Universalist Ministry for Earth, Quaker Earthcare Witness, Web of Creation. Other organizations cater to multiple types of bodies within the same tradition. Hazon, the Shalom Center, the Baltimore Jewish Environmental Network, and the Coalition on the Environment and Jewish Life (COEJL) speak to the four major Jewish denominations in the United States; Restoring Eden, A Rocha, the Southern Baptist Climate, and the Evangelical Environmental Network operate as pan-Evangelical organizations; while the National Catholic Rural Life Conference (NCRLC) and the Catholic Coalition on Climate Change (CCCC) work in relationship with the American Catholic Church.

The CCCC explicitly ties its mission to the larger agenda of the United States Conference of Catholic Bishops and its teachings about prudence, poverty, and the common good. Quaker Earthcare Witness (QEW) claims its mission is rooted in the "Historic Quaker principle of 'seeking Truth together' and in the goal of 'living in a right relationship with all of creation.'" Hollister Knowlton, former clerk of QEW, notes that "'Right relationship' has a real spiritual weight to it. For us, it means understanding that we are all one; that we are a part of the earth. And when we live in a right relationship with the earth, we are no longer in an exploitative relationship with the earth." Hazon pushes an ambitious agenda to renew and reenergize all of Judaism, in part by tapping into and developing a centuries-old tradition:

> Our vision is to create a healthier and more sustainable Jewish community— as a step towards a healthier and more sustainable world for all. Our vision is of a renewed Jewish community: one that is rooted in Jewish tradition, engaged with the world around us, radically inclusive, passionate, and creative. . . . Our vision is of a community that engages vitally with the world in which we live, and in doing so achieves two things: the renewal of Jewish life itself and the addition of a further chapter in the distinctive contribution that Jewish people have made to the world for 3,000 years.

REMOs then justify their mission to awaken religious groups to environmental action by grounding it in theological, scriptural, and ethical ideals of specific traditions. Doing so demonstrates their knowledge of a tradition or traditions and signals to their religious audiences that they belong to that community and should be considered to be legitimate actors. In many ways, REMOs are placing themselves in a prophetic role, and like their models from the Old and New Testaments, their job is not simply to awaken the religious community, but to prod it into action. The prophetic call often urges the people to return to a more careful observance of religious teachings, and in the case of REMOs, it is a call to recognize that one's religion has always included a green mandate.

Mobilizing Religious Networks

As suggested by several quotes from the previous section, many REMOs aim to create networks of individuals, congregations, or larger religious bodies that will then take practical steps toward alleviating some set of pressing ecological concerns. Implicitly, this goal reflects the need to mobilize and harness the collective power of religious communities in order to have any impact on complex environmental problems. For example, the newly formed Chesapeake Covenant Congregations says its mission is to: "create and nurture a network of churches in the Chesapeake watershed in covenant with God to establish healing ministries for lands and waters, to revere and cherish the earth, and restore the ability of the Chesapeake Bay to sustain all life."[22] Most REMOs are embedded in religiously homogeneous networks. Liberal Protestants tend to recruit other liberal Protestants; evangelical REMOs mobilize other evangelicals. REMOs serving a single denomination or tradition commonly tried to mobilize congregations, judicatories, and/or clergy associations within that tradition. Many REMOs had few or only weak ties to organizations and individuals outside of their home tradition. For example, REMOs with strong ties to liberal religious groups have been unable to connect with evangelical and African American religious communities; all of the Jewish groups have been unable to mobilize Orthodox congregations or communities; and apart from Faith in Place, no organization in the study has made serious efforts to mobilize Islamic congregations or religious bodies.

Thus REMOs' embeddedness in specific religious networks and in specific religious traditions jointly shapes how they develop their mission and organizational identities. Some religious groups have rules about who are appropriate partners for conversation and joint activities, and may be able to sanction REMOs who violate those rules. For example, both Catholic REMOs note that the church hierarchy would not allow them to work closely and/or publicly with organizations that did not support the church's pro-life positions. Thus both the CCCC and the NCRLC would not be able to work with or recruit from groups that advocated for the use of contraception as a means of controlling population growth. Evangelical REMOs tend to face similar restrictions with sanctions that are primarily informal, as suggested by Jonathan Merritt's experiences with the resistance from the Southern Baptist Conference. REMOS that emerge from liberal and mainline Protestant and/or Reform Judaism have rules about inclusivity and cooperation that shape how they go about acting on their mission to awaken and mobilize religions (this is discussed in greater detail in chapter 5). Interfaith and ecumenical REMOs also must develop messages and practices to effectively address religious diversity. On the one hand they must demonstrate sensitivity to the theological particularities of each tradition in the coalition while also creating a common view of environmentalism that will cut across denominational and religious lines and thus foster joint action. This balancing act is evident in the following description of how Georgia Interfaith Power and Light operates:

> We really try to be specific to that tradition and talk about things from their tradition, from their scriptures, that would speak to them. And not just

broadly Christian traditions, for instance, but also denominationally—do they have anything in their denomination that would speak to them? But there are common themes that seem to come up and resonate in all of these traditions, in one way or another. So stewardship is one; justice one; interconnectedness; and awe and wonder.

Even REMOs that serve a single denomination or a tradition must attend to internal diversity in order to be seen as credible representatives of their tradition and mobilize support, as illustrated in the history of the Coalition on the Environment and Jewish Life (COEJL). Leaders of COEJL had strong ties to leaders of twenty-nine national Jewish bodies, and their embeddedness in this network played two crucial roles in the development of COEJL's mission and identity. First, one of COEJL's founding partners was the Jewish Theological Seminary of America, and its scholars led the effort to develop the distinctly religious mission for the new organization during its early years. The seminary helped COEJL develop a theological message that could demonstrate how environmentalism was integral to Jewish identity and practice. They anchored COEJL's mission in several core values that cut across Judaism that sent clear signals to a multidenominational audience that Jewish environmentalism was justified, appropriate, and valid. Such claims are evinced in the following from COEJL's ten-year anniversary report:

> COEJL seeks to expand the contemporary understandings of such Jewish values as *tikkun olam* (repairing the world) and *tzedek* (justice) to include the protection of both people and other species from environmental degradation. COEJL seeks to extend such traditions as social action and *g'milut hasadim* (performing deeds of loving kindness) to environmental action and advocacy. And *shalom* (peace or wholeness), which is at the very core of Jewish aspirations, is in its full sense harmony in all creation [italics original].[23]

Second, this network of national Jewish organizations pushed the founders of COEJL to adopt a national focus for its work. In its founding period, COEJL worked closely with two of the national Jewish policy organizations, the Jewish Council for Public Affairs and the Religious Action Center for Reform Judaism, to develop policy positions and advocacy efforts on national legislative issues, such as the 1996 campaign to save the Endangered Species Act.

In short, COEJL established its legitimacy by mobilizing important leaders across a broad spectrum of American Judaism and then allowed those leaders to develop and promulgate the new REMO's mission. More generally, REMOs that were strongly embedded in particular religious communities tended to develop organizational missions that were religiously distinct, even sectarian, because key stakeholders (for example, bishops, important denominational leaders, or clergy) could withhold support or actively oppose the new REMOs. Thus tradition and networks work together as they direct REMO leaders to create their mission and goals. The former places the boundaries around the appropriate content of mission statements, and the latter help REMOs identify their audience, enforce the boundaries of appropriateness, and provide legitimacy by endorsing REMOs.

Acting Locally

The third widely shared mission among REMOs was to focus on local or regional issues and populations. They seemed to have taken the adage "Think Globally; Act Locally" to heart. Just over one-quarter of the sixty-three groups in the study have a national organization or a national focus and agenda. However, many of the national groups have state-based chapters or support state and regional activities. For example, the National Council of Churches (NCC) represents national religious voices on public policy issues at the federal level (for example, working on the Farm Bill), while at the same time it sponsored state-level programs such as the interfaith global climate change initiative in the late 1990s and early 2000s, and its current program on environmental health works with parachurch groups in Minnesota, Michigan, Massachusetts, Maine, and Washington. Similarly, the Regeneration Project (RP) aims to establish and support semiautonomous state-based Interfaith Power and Light (IPL) organizations, while at the same time engage in advocacy and public policy discussions about energy and climate change at the national level. This local and national orientation is evident in the mission statement from the Regeneration Project's website:

> Interfaith Power and Light is working to help congregations be models of energy efficiency, utilize renewable energy, and to lead by showing a strong example of stewardship of creation. At the same time, we know that we cannot stem the tide of global warming by our actions alone, and therefore we actively support public policies to reduce society-wide U.S. emissions to a sustainable level.

The majority of the groups in the study identify specific geographical boundaries within which they work. Earth Ministry (EM) serves the Puget

Sound watershed; Earth Sangha works in northern Virginia; GreenFaith educates and advocates in New Jersey; Voices for Earth Justice and Partnership for Earth Spirituality operate in Michigan and New Mexico, respectively; LEAF and Christians for the Mountains work to preserve the mountains of Appalachia. Several groups have even narrower geographical boundaries—Los Angeles for Progressive Christians Uniting's Eighth Day Project, Atlanta for Earth Covenant Ministry, or the Detroit River for the Grosse Ile Congregations group. Many groups identify rootedness in place as an important value. Some organizations, like the Maine Council of Churches (MCC), speak more generally about their commitment to work locally, as evident in the council's statement about its focus of "helping congregations build sustainable community through practicing environmental and economic justice in our own neighborhoods and beyond." Andy Burt, the Maine Council of Churches environmental program director, elaborated on this commitment during our interview. She described their Good Apple program in which congregations pledge to support local farms and sustainable agricultural practices. Congregations do not simply make a financial pledge but enter into a covenant with local farmers in order to build community. Similarly, Hazon has partnered with local farmers in sixteen states to create sixty CSAs (Community Supported Agriculture groups), and the Interfaith Network for Earth Concerns in Oregon has a program to connect congregations in the state with farmers through CSAs, bringing famers to congregations to sell their produce after Sunday services, and organizing congregational buying clubs.

While REMOs' embeddedness in specific eco-spaces is partly responsible for their commitment to the local, state, or regional orientation of most faith-based environmentalism, it alone is not a complete explanation, especially since environmental problems often expand beyond precise geographic boundaries. The mission to act locally also reflects a strategic response to constraints imposed by resources and constituencies. Few organizations possess the financial and human resources to operate at the national level. Few groups have more than a handful of staff members, and many operate simply with an executive director and a board. Several groups with green congregation programs noted that they lacked the capacity to track and offer ongoing education or support to congregations in these programs. Seventy percent of the organizations have annual budgets less than $250,000 (54 percent are less than $100,000), which puts limits on the scope of their activities. Moreover, most of the founders' networks were local—rooted in city or state clergy associations and congregations—which put the brakes on national aspirations for most REMOs.

Nearly every organization in the study identifies its primary constituency as communities of faith, often congregations. Some groups even identify their primary constituency as congregations or religious bodies within a particular geographic location. For example, Michigan Interfaith Power and Light identifies itself as "coalition of over 100 congregations across the state of Michigan, whose mission is to involve communities of faith as stewards of God's creation in promoting and implementing energy conservation, energy efficiency, renewable energy, and related-sustainable practices."[24] By emphasizing local environmental issues, REMOs fit into the norms and practices about social action that are familiar to their audiences, since many congregations also focus their own acts of justice, charity, or advocacy at the local level.[25] An approach that is sensitive to the existing ways that congregations work may be an important way to gain legitimacy and thus ultimately to realize the larger goal of awakening congregations to the religious call for environmental action. REMOs adopt a local mission in part because it fits into the identity and culture of the congregations they are trying to mobilize. By offering resources that mesh with or complement congregations' sense of "who we are" and "how we do things here," REMO leaders hope to demonstrate the legitimacy of religious environmentalism.[26] At the same time, this local strategy may create new problems for the nascent movement insofar as it pushes REMOs away from creating the kind of national, federated organizations that would make it easier to coordinate messages and collective action.[27] Thus embeddedness in local religious fields may make REMOs appear less legitimate because they cannot engage in the kind of activities performed by more mature movement organizations or mount the kinds of polished and skillful campaigns that secular environmental groups mount.

Niche Activism

REMOs' concerns about legitimacy and their limited resources compel them to devise a set of activities that will appeal to their potential supporters and be financially sustainable over time. To this end, most REMOs engage in "niche activism." That is, organizations tend to work on a particular issue or set of issues and rely primarily on a handful of mobilization and protest strategies to realize their goals. In her study of the GLBQT movement in Chicago, Levitsky found that many organizations in the movement "developed 'expertise' in specific tasks—litigation, lobbying, protest, consciousness raising, direct services—that they then deployed as part of a multipronged social reform effort."[28] She suggests that niche activism can promote organizational interdependence and cooperation and therefore can be a powerful way to marshal a fuller set of resources for collective action than would otherwise be available. This finding holds true to some extent for the REMOs in this study, especially in terms of actively seeking to specialize in certain activities and promoting intramovement cooperation rather than competition.

Many organizations identify a dominant issue or small set of issues on which they work. Some REMOs downplay specific ecological issues and instead emphasize particular tactics. For example, some organizations see their mission primarily as one in which they educate individuals and religious groups to raise awareness and mobilize practical action (like making a congregation energy efficient), while others see themselves as a provider of resources and information and do little advocacy or community building. Levitsky's attention to niche activism also points to the important role cooperation plays within social movement fields, and how movement organizations make the best use of their particular resources. REMOs commonly borrow the resources and ideas from one another or refer their supporters to other REMOs as they see no need to duplicate the mission, and few possess the wherewithal to be full-service nonprofits. For example, the Baltimore Jewish Environmental Network (BJEN) focuses its mission on helping local congregations by creating a green congregation handbook. Rather than write it from scratch, its author confessed that she "shamelessly raided all the Internet information sites I could find. I said, 'Guys, we do not need to waste our time reinventing the wheel. Let's just see what's out there,' and it was everything from what the Universalists/Unitarians are doing because they do great stuff, [to] things on the Catholic websites, as well as COEJL's website. So anyway, we put together a 10-page document or so, which focused on four areas, and that is the template for our greening synagogues work [at] BJEN." Many interviewees report that they use resources (or post links on their web pages) from Earth Ministry (EM), the National Religious Partnership for the Environment, the National Council of Churches, or the Regeneration Project, rather than create their own educational or greening materials. Renee Rico, executive director of Presbyterians for Restoring Creation (PRC), defends the PRC's focus on mobilizing the denomination and educational goals rather than engaging in the kind of advocacy done by the NCC or the Regeneration Project: "They're the experts in doing that kind of advocacy. We're not trying to duplicate what they do because you need staff; you need to understand the legislation and all of that. We're not big enough to do that yet." Thus cooperation and specialization characterize the kind of activism in which REMOs engage.

Developing Issue Niches

Four main issue niches emerged from the analysis of interviews and written materials: global warming and energy; environmental health and toxins; sustainable agriculture and sacred food; and a fourth niche that emphasized the natural world around such issues as preserving biodiversity, watersheds, forests, or combating air and water pollution. Given the growing importance of global warming, it was not surprising to find nearly 90 percent of all REMOs worked on this issue and about one-quarter were dedicated to this issue alone. Cassandra Carmichael provides an explanation for why most organizations place resources and attention on global warming:

> Well, we have our bread-and-butter, main, core issue—climate and energy—and that's because it's the global issue. It trumps every other issue. . . . You can't really address biodiversity and forget about climate because, it's just going to come in and squash it, and because also, it's a huge moral issue from the church standpoint.

Another quarter of the REMOs focus on global warming and biodiversity issues, especially protecting watersheds and forests. Roughly 10 percent of REMOs worked on environmental health issues and another 10 percent emphasized sustainable agriculture and how buying local and/or organic can lower carbon emissions associated with the industrial food system. Only five REMOs worked on all four issues.

REMOs' move toward specialization is driven by a concern to avoid duplicating the services and practices of their peers and thus to maximize their resources. This is particularly evident with REMOs that work in the same field as one of the thirty-nine state Interfaith Power and Light (IPL) organizations. These REMOs adopt one of the other issues in addition to climate change since they do not need to offer congregational greening projects or energy audits and can offer something unique that the IPLs do not provide. For example, the Eco-Justice Collaborative works on climate change, but to distinguish itself from Faith in Place (which doubles as Illinois Power and Light), it also works with community and environmental nonprofit groups in Chicago's south side to remedy toxic waste and promote environmental health initiatives.

Some REMOs focus on the issues that are most salient to their targeted populations (for example, environmental health for Jesus People against Pollution (JPAP) and GreenFaith audiences who live in the toxic landscapes of Mississippi and New Jersey, respectively), while others create issue niches that draw on the religious values that animate their communities. That is, REMOs select a set of issues that they believe can be tied to specific scriptural or religious mandates, are important to their audience, and can easily be justified to congregations and religious individuals. This strategy is evident in the following appeal to oppose mountaintop removal offered by Christians for the Mountain (CFM):

> God has made us His stewards, and we must give account for our stewardship to our Creator. We may not avoid this responsibility, for it is God-given. To poison, pollute, and destroy God's Creation is sinful. When we do this, we heap contempt on our Creator, and we endanger our health, our neighbor's health, and the health of future generations. We endanger the welfare and survival of God's work.[29]

Here CFM couches its appeal in the language of stewardship, sin, divine judgment, and the relationship of the believer to God, all of which should resonate with the conservative Christian audience it hopes to mobilize. The group also takes an issue about nature and makes it an issue about human health in order to cater to the human-centered orientation toward social issues its audience favors. Organizations that work on biodiversity issues also frame them as essentially about human beings, especially the human-divine relationship, in part because there are few directives from religious traditions about caring for animals or nature. REMOs that work on food issues often justify their work by arguing that eating and food production are fundamentally moral acts. The former interim executive director of the National Catholic Rural Life Conference (NCRLC) claims, "Where our food comes from, how it is produced, the welfare of animals, the impact of practices upon the environment, the availability of food for the poor, the common-good" are the guiding principles for its food and farm programs. On its web page the NCRLC argues, "Agriculture is not only how we grow food, but how we treat those who bring food to our tables and about how we make sure the whole human family is fed." These statements key into important Catholic values about justice for the poor, human dignity, the sacredness of creation, solidarity, and stewardship.[30] REMOs, then, create issue niches that help them avoid competing with and duplicating the work of other REMOs, that resonate with the lived experiences and ecological problems facing their constituents, and that can be framed as fundamentally religious rather than as ecological, technical, or scientific issues. Thus REMOs' embeddedness in particular religious cultures and audiences shapes how they go about establishing issue niches. Climate change and environmental health are easy niches to create because REMOs can tie these issues to powerful religious mandates to help human beings. However, they struggle to justify programs to preserve species and land (only 40 percent of all REMOs work on these issues), but do so by framing them as actions that ultimately benefit humans or deepen one's relationship with God.

Crafting Strategic Niches

REMOs also set up strategic niches in which they emphasize a primary orientation toward mobilization and protest along with a small set of activities (that is, a repertoire of contention) that will help them realize their mission.[31] Like their secular counterparts, REMOs most commonly rely on the strategy of promoting public awareness or consciousness-raising through education; however, they focus on a different audience (religious individuals, congregations, and other bodies rather than the media or political actors) and use different tactics (for example, REMOs are far more likely to offer green curricula for congregations or resources for greening a religious facility than to make presentations to or work with government agencies as EMOs do).[32] Most organizations rely on several strategies, although one is often given greater attention or importance depending on organizational resources and interests of staff, boards, and/or constituents. REMOs appear to be known for certain kinds of activities or for providing particular kinds of resources. The state IPLs are the REMOs to contact if one's congregation wants to learn about how to "green" their facilities, while GreenFaith, Eco-Justice Ministries, Web of Creation, or Earth Ministry are the first stops for educational materials.

The most common strategies (in order of importance) include promoting public awareness or consciousness-raising through education, providing resources, cultivating spirituality, modeling and promoting sustainable lifestyles (prefigurative action),[33] engaging in advocacy, organizing (building interorganizational networks or serving as an incubator or catalyst for other groups), and promoting direct action (for example, protest or litigation, conservation). Many groups bundle several strategies together, while a small number specialize in one to two strategies. Several groups used the metaphor of a three- or four-legged stool to describe their bundle of strategies or strategic agenda. For example, at Earth Ministry in Seattle, when I asked about the organization's primary goal, I was told by staff member Deanna Matzen that they operate with a three-legged stool: "The first leg of the stool is education. . . . The second leg is community building and institutional actions. And that would be our congregational programs, like the greening congregations program and working with churches and religious institutions. And then the third leg is faith-based advocacy and systemic change." Similarly, Patricia Gillis, the executive director of Voices for Earth Justice, noted that they operate with a four-legged stool of prayer, study, action, and community. The New Community Project's mission is to promote stewardship, peace through justice, and experiential learning, which suggests that direct action and education are its two strategies.

Nearly all of the organizations rely on education as a main strategy, and so there really isn't an environmental education niche per se (about 10 percent of the REMOs only offer environmental courses or workshops for religious groups and individuals). Some offer courses or materials about the issues (for example, state IPLs often provide information in their courses to congregations about the science behind global warming); some offer more explicit religious education (usually about what traditions teach about the environment) that may help individuals and congregations cultivate a green ethic and/or a green spirituality; still others offer experiential education by providing tours of toxic sites (for example, GreenFaith), trips to meet groups profoundly affected by global warming and environmental degradation (Sacred Earth Network, New Community Project), retreats or workshops to cultivate spirituality (Sacred Earth Network, Earth and Spirit Council, Partnership for Earth Spirituality), or outdoor excursions (Hazon's Jewish Environmental Bike Rides). The REMOs that focus on global warming and energy tend to offer practical education (how to make facilities energy efficient). In all of these endeavors REMOs' tight embeddedness in particular religious and ecological communities shape their consciousness-raising endeavors.

Despite these different approaches to education, all the groups attempt to connect particular goals with the theological values and cultures of their target audiences. Katy Hinman from Georgia Interfaith Power and Light notes how they provide straightforward environmental education surrounding the causes of climate change as well as theological education. They most often offer classes or presentations that help congregations connect climate change and their tradition:

A lot of what we do is that kind of theological education piece—how does this fit into my faith tradition? What does scripture have to say about caring for the earth? Why is this something we're talking about in my faith community? Because there are plenty of arguments—economic and environmental arguments—but not many that make that faith connection. So then we also provide resources for congregations to start incorporating some of these themes into their worship. A lot of times it's just directing people to resources they already have. I mean there's [sic] tons of hymns, for instance, that have these themes, and people just don't think about. So just opening peoples' consciousness to be aware of that. . . . And also providing, new resources that they may not have, or helping them write some prayers that they may want to use.

In this statement, Hinman shows how they go about acting on their mission of making environmentalism an integral part of the life of faith and awakening people to the sacred call to care for the earth. She notes that they do so by customizing their programs to the specific religious traditions of their audiences. Recall that REMOs are embedded in specific audiences who think about the environment, religious ethics, and the nature of sacred texts in particular ways. REMOs attempt to tune their educational programs to their specific audiences' theological and ethical sensibilities and are aware that the failure to do so will damage their legitimacy and efficacy. This concern to work within and not against a given tradition is evident in the discussion I had with the leader of Earthcare:

I think we are pretty much solidly, traditionally Christians. One of the issues that we've had to deal with in planning our conferences and our retreats is understanding what constituency we're trying to speak to, and try and stay pretty middle of the road theologically, just because our group is as diverse as it is. When somebody hears about our speaker, they think we might want to involve, we have to read something they've written or make sure they're not too off the wall theologically, for the Southern Baptists in the group or the fundamentalists or the Episcopalian and the Lutherans.

Roughly 30 percent of all REMOs occupy a niche that combines education and resource provision. As suggested above, the resources are often quite varied. Most commonly, REMOs provide congregations with help greening their facilities or their individual homes. They also routinely list other organizations' websites where people browsing the Internet can learn about specific issues, or find like-minded activists. Some groups offer congregation-based environmental handbooks or environmental curricula (Earth Ministry, Unitarian Universalist Ministry for Earth, the Web of Creation) or links to such curricula (for example, Eco-Justice Ministries, Presbyterians for Restoring Creation). A few organizations, such as the Network Alliance of Congregations Caring for the Earth (NACCE), Web of Creation, Eco-Justice Ministries, the Orange County Interfaith Coalition for the Environment, and Earthcare are resource specialists. They provide curricula, pamphlets, and handbooks, and they often organize conferences as a way of promoting information and resources on specific issues.

Another 30 percent follow some kind of prefigurative strategy in which they provide resources for individuals and religious bodies to live in more sustainable ways or urge their audiences to challenge the consumption-driven lifestyle of contemporary American capitalism. They offer models for enacting or embodying a religious green ethic. For example, the Eco-Justice Collaborative in Chicago criticizes capitalism and aims to make people reflect on their marketplace choices. In the discussion of its mission on its web page, EJC's founders discuss why they challenge Western materialism and consumerism and illustrate how they integrate religious and ecological values:

> Our mission is to raise public awareness of the consequences of lifestyle choices on people and our planet and to encourage changes that seek harmony with Creation; respect all life; value diversity; support ecological sustainability; and bring about a just distribution of the world's resources. . . . Within just a few generations, rapidly expanding population, world-wide industrialization, astounding technological advances, and the concentration of economic and political power have created social and environmental issues of life and death proportions. Issues that could be ignored in the past are now ecological and moral imperatives. . . . We can no longer turn away as if tomorrow's technologies will reverse trends that have been set in motion. If we are going to prevent the loss of biodiversity, lessen the effects of global warming, and eliminate the plague of persistent poverty, we need to transform our ways of thinking and acting now. This includes understanding how our lifestyle and consumer choices affect the planet, and then being willing to choose to live differently so that we lessen those impacts.

This statement draws on a set of pan-religious and deeply held values (for example, stewardship, justice) that can appeal to a diverse religious audience and ties those values to proenvironmental activities. It also illustrates the central role education and consciousness-raising play and how groups in this niche combine these strategies with prefigurative action.

About 15 percent of REMOs yoke advocacy and organizing as their central tasks. In general these are the national REMOs who have more staff, more financial resources, and a national orientation and/or well-developed activist or organizational networks. They see the scope and scale of the environmental problems facing the world and believe solutions must be forged via national legislation. For example, the NCRLC works on the Farm

Bill, and the NRPE, Regeneration Project (RP), and the NCC have lobbied for climate change legislation in Congress. They also are the organizations that provide start-up funding for other REMOs and develop programs that employ other REMOs (for example, the NCC created and funded several state interfaith global climate change organizations and an environmental health program that several REMOs in the study operate on behalf of the NCC). Thus a handful of REMOs occupy a niche as movement midwives or catalysts and are largely the public face of the new movement. Another 20 percent of all REMOs also engage in advocacy in a limited manner. Most focus on state-level issues, and generally they attend the annual religious lobby days in state capitals or work with both faith-based and secular organizations on specific legislation. However, for these REMOs advocacy is not their primary activity, and their 501c3 status limits their ability to lobby or engage in advocacy.

Finally, a handful of organizations (about 10 percent) engage in direct action such as protests (Religious Witness for the Earth), litigation (Green-Faith, Partnership for Earth Spirituality), or activities to preserve natural areas and/or curb pollution (Earth Sangha, Green Sangha, Floresta, and A Rocha, USA) while also promoting some form of eco-spirituality. Green Sangha (GS) typically combines meditation and such activities as habitation restoration or letter-writing campaigns to eliminate plastic bags. The Partnership for Earth Spirituality (PES) offers spirituality retreats (as part of its Catholic roots) and joins with other New Mexico environmental groups protesting and litigating against nuclear energy. Both claim that their actions derive from or are driven by the groups' spiritual practices. For GS, meditation cultivates an awareness of the interdependence of humans and nature, love, and forgiveness, which the REMO then encourages its members to act on in concrete ways. According to Sister Joan Brown, the leader of PES, their contemplative work ("listening to the earth, trying to come to sensitivity within ourselves, and seek[ing] the wisdom and guidance of the Creator") leads toward "advocacy and other actions." In both examples the REMOs harness the ideals and practices from particular traditions and show their audiences how their own tradition must compel them to take proenvironmental action. They gain legitimacy by emphasizing the spiritual over the environmental or by explaining how sacred texts and historic traditions encourage and ground environmental politics.

Religious Witness for the Earth (RWE) is unique among REMOs insofar as it is the only organization to engage in direct protest and political demonstrations. Their use of the term "witness" carries distinct Christian meaning and refers to believers' efforts to explain or demonstrate beliefs or to tell the story of Jesus so that others may come to believe as well. The Reverend Fred Small notes that some have criticized RWE because witnessing is passive and not very effective in realizing specific environmental goals. But he explains that the believer's job is not to achieve some specific result but to demonstrate in support of certain truth claims. Small doesn't see witnessing as powerless. Rather, he argues that it is a mode of empowering people and communities of faith as suggested in his description of why RWE sponsored a walk across Boston to protest government inaction on climate change:

> You pray, and then you act. You pray, and then you demonstrate. We were taking our prayer out of the sanctuary and into the street or into the Public Square. And that's, I think, what is special about Religious Witness for the Earth. We bridge those two worlds. We're not just about individual responsibility; we're not just about religious leaders issuing joint statements of grave concern; we're not just about children's religious education curriculum; we're not just about green congregations or congregational buildings. All of these are important, but we say we must bear witness; we must leave the safety of our sanctuaries and speak with authority and courage in the public domain.

Small also notes how RWE is less interested in achieving specific policy and legislative ends and more interested in enriching the religious lives of those who participate in their events:

> But that doesn't mean that we don't care about results; it just means that we're not attached to them. We absolutely want to be strategic; we want to be effective. But we're not a lobbying organization; we don't do head counts on work, you know, how many votes are we short of this particular bill. . . . But there's no question in my mind that the event is going to make a difference.

In Small's view ultimately God takes care of the results, and RWE's mission simply is to be a prophetic voice. Thus RWE makes its appeal for legitimacy by showing how protest is driven by religious values, and its first goal is religious transformation rather than saving the environment. This distances it from secular movement organizations and provides religious groups and individuals a powerful reason to get involved.

Conclusion: Embedded Entrepreneurship

The leaders of these new movement organizations face a number of challenging tasks in order to gain a foothold within American religion. The organizations' literature stresses the central role legitimacy plays in this task and argues that institutional entrepreneurs must act strategically and carefully to successfully establish new organizations or organizational practices. I have argued that REMOs' leaders' knowledge of the religious and environmental movement fields helped them carefully craft their organizational missions and activities to gain the requisite institutional legitimacy.

First, REMO leaders needed to overcome the indifference and suspicion about environmentalism that many in the religious field felt. They had to demonstrate that environmentalism is not inimical to the life of faith but integral to it. They did so by connecting their ecological work to specific religions' core values and theological traditions, and by using religious rather than secular environmental language (for example, "care of creation" rather than "preserving nature"). Thus they demonstrated that this nascent movement conformed to the extant and operative beliefs, meanings, and rules of religious bodies. At the same time they introduced important innovations in religious ethics by demonstrating how love of neighbor, stewardship, or acts of justice and compassion must include the environment if the life of faith is to be authentic and efficacious. But REMOs needed to be mindful that their potential audiences and supporters would not label them as merely an environmental organization, or, as Paul Gorman wryly noted in chapter 1, that they could not be seen as "the environmental movement in prayer." They used their justification of their missions, goals, and activities to draw boundaries between themselves and the secular movement and as a way of defining their work as distinctly religious and hence more legitimate.

This process of organizational innovation mirrors the kind of change in worship that Chaves documents insofar as REMOs "reshuffle" or play with elements of the tradition, thereby ensuring continuity with the old, while also introducing new ideas. Doing so signals to their religious audiences that REMOs are "recognizably similar to and recognizably different" from existing religions and environmental organizations.[34] This ability to use and modify elements from different religious traditions or cultures was a result of REMOs' strong embeddedness in particular religious communities, and doing so was an important way of establishing their religious credentials and the identity of the new movement. Thus innovation in organizational form and practice emerged from the center rather than the periphery as the literature suggests.

Second, REMOs strategic use of religious traditions also helped them create a collective identity that was easy for their audiences to adopt since it was consistent with their existing religious identities. They did not require audience members to abandon their old religious identities. One did not have to radically change what it meant to be a good Presbyterian, Jew, or Buddhist, but only had to broaden this identity to include some commitment to saving the environment. REMOs argued that becoming a religious environmentalist was a critical way of becoming a more authentic and complete believer. REMO leaders relied on their knowledge of denominational leaders, clergy, and rank-and-file members to anticipate how they would respond to their efforts at creating religious environmentalism. The distinctly religious language and goals of the new movement's organizations reflect their attempts to avoid alienating their real and potential supporters. REMOs that adopted an interfaith or ecumenical orientation wisely emphasized common values and ideals that cut across religions or adopted the "walk together separately" strategy to avoid forcing supporters from different traditions to compromise on their commitments. Their sensitivity to audiences' perceptions, values, and religious sensibilities about environmentalism was an important way of gaining trust and support. In the next chapter, I expand on the story of how REMOs created green religious traditions and, in particular, pay special attention to the role audiences play in the formation of new green religious culture.

Third, REMOs' leaders' embeddedness in specific interpersonal and organizational networks played an important role in the development of their organizational missions, identities, and goals. REMOs rely on members of their network to help them understand the nuances of specific traditions and how to make their message congruent with a given tradition. This practice is best illustrated in the foundational work of the NRPE and Paul Gorman's concern that the first step was to get the theology right. Members of their networks also are the gatekeepers to particular religious communities, and their endorsement of a REMO can enhance their credibility within and improve their access to such communities. In brief, cultural and network embeddedness work hand in hand to help REMOs launch and establish their fledging organizations.

Fourth, REMOs' strong embeddedness in particular religious fields provided them with a deep understanding of the limited availability of resources from denominations, judicatories, and congregations. They made a number of organizational and tactical decisions to work around the resource constraints they faced. REMOs tend to work on local, state, or regional issues, this both keeps costs down and has the virtue of demonstrating their relevance by addressing concerns that touch their potential supporters more directly.[35] Most REMOs have adopted a coalitional form and thus have a deeper pool of potential donors and members. REMOs keep their costs down by relying heavily on volunteers and maintaining small staffs.

Finally, they have created specialized niches in order to avoid duplicating the programs and projects of other REMOs, borrowed resources from one another, and often referred their audiences to other REMOs when they could not offer the information or programs their audiences desired. Thus rather than engage in the kind of competitive, zero-sum political game that neo-institutional scholars of organizational emergence describe, REMOs follow a more cooperative politics of persuasion.

In this chapter I have tried to heed Jasper's call for a more "strategic" perspective in the study of social movements by highlighting the "complex cultural processes" through which the new movement organizations attempt to stake a place in the movement and religious fields, garner legitimacy and support, and mobilize action.[36] I have shown how REMOs' choices about organizational mission and form, issues, and tactics are constrained, channeled, and facilitated by their embeddedness in specific religious organizational networks and traditions, and, as will be more evident in the next chapter, how their organizing was shaped by the responses they expected to receive from the stakeholders and audience members in their particular religious fields.

FOUR

Creating Religious Environmental Traditions

Rev. Peter Sawtell, the executive director of Eco-Justice Ministries (EJM), is well aware of how difficult it is to persuade his fellow Christians to become environmentalists. EJM is an ecumenical REMO with a national audience of mainline and liberal Protestants. Three years earlier Sawtell had helped organize a meeting with nine other REMOs when they all realized that mainline Protestant denominations were curtailing programs and staffing for environmental issues due to budget cuts. They decided that "if there was going to be a strong push into congregations and communities, it was going to have to come from independent agencies." But the challenges are not simply organizational. They are perceptual; they are rooted in how Christians understand the purpose of the church, read and interpret scripture, and place boundaries around what constitutes legitimate religious social issues. Sawtell notes:

> We need to look at how we relate to the natural world, how do we deal with the justice questions. The other big thing that I've realized needs to be said very explicitly is that churches need to think of themselves as transformational. One reason the environmental stuff just isn't getting traction in an awful lot of local churches is that they really see their role to be fairly quiet and pastoral, [a] "help people feel good and fit in" perspective. Before we can talk about environmental change, we have to get churches thinking about themselves as change agents at all.

As we talked more about the challenges facing EJM and other REMOs, Sawtell directed me to several of the newsletters archived on EJM's website to learn more about his diagnosis of the problems. One essay about a workshop on the Bible and the environment not only points to the perceptual and legitimation challenges facing REMOs, but also suggests how they are trying to address them. He begins the essay by noting how difficult it is for some to make sense of "disoriented, jumbled, or confused" biblical passages, especially when today's readers do not understand the context and intentions of the compilers of the Bible. As a result, he argues, we tend to miss the full meaning of many texts. In particular, his workshop participants initially missed how the passage from Exodus that they were studying contains instructions about providing justice for the nonhuman parts of the community. He concludes the essay by urging Christians to reimagine how they think about faith and the environment:

> During Tuesday's workshop session, one person made a comment about how "eco-justice" is not a big thing in the Bible. . . . The Bible may not have a lot of passages giving explicit definition to what we think of as eco-justice. At that level, her comment was accurate. But at a deeper level, large sections of scripture carry assumptions and values which are foundational for our eco-justice ethic. When we come to the Bible with our eyes, ears, and hearts open to a larger sense of community, and a more encompassing sense of justice, we discover a rich and faithful grounding for addressing today's most urgent issues. . . . From the first chapter of Genesis to the last chapter of Revelation, the biblical witness consistently and frequently affirms God's care for creation and each creature, especially the most vulnerable—both human and nonhuman.[1]

Sawtell's comments suggest that religious environmentalists face a difficult organizing problem because their specific traditions and sacred texts can appear to be relatively silent on ecological issues. This means that individuals and congregations may struggle to see how faith and the environment are connected or may not even imagine the two can be connected. Thus REMO leaders must persuade religious individuals and organizations that holding an environmental religious identity and engaging in environmental action are consistent with and even mandated by religion. The essay about the Bible quoted above also highlights an additional challenge for REMOs: religions' primary concern about human salvation, suffering, and flourishing may blind religious groups and individuals to environmental problems or make them suspicious of working on issues that could pull them away from human-centered issues. Sawtell is especially concerned about the task of yoking faith and environmental action, and he notes that EJM's aim is to provide the stories, ideas, and language that resonate with religious audiences. When I asked Sawtell how he tries to sell the environment to skeptical or unaware audiences, he notes that:

> Years ago when somebody asked what key text I would come back to, I said, "Love God and love your neighbor," which is not one of the environmental texts. Then we have to expand the definition of neighbor. That's the very, very short answer. More significantly, I keep coming back again and again to the Hebrew principle of shalom as I put it—peace, justice, and harmony for all of God's creation. That meshes very closely with how some people have defined the theological ethical principle of eco-justice, which is the well-being of all humanity on a thriving earth. So I would come back to shalom and eco-justice and love of neighbor as sort of the grounding principles that I refer to over and over and over.

At the most general level, Sawtell is engaged in the process of reframing or reinterpreting religious narratives, images, and symbols to make them applicable to ecological issues. More specifically, he relies on "frame bridging" insofar as he attempts to connect environmentalism with an existing, but nonecological, human-centered ethic of care of one's neighbor and "bricolage," in which he combines Jewish and Christian ideals to form a new ethic.[2] The focal question of this chapter is, "How do activists create a religious environmental ethic, a new green religious identity, and a set of green religious traditions?" These are difficult tasks that go beyond the usual challenges of frame alignment because few religions have existing environmental ethics, theologies, or a history of concern with nature. Moreover, the indifference and suspicion about environmentalism within many religious communities exacerbate the challenge of creating a green religious tradition. At the same time, this empirical question raises theoretical questions about the nature of cultural creativity and innovation: how do activists borrow, adapt, and rework resources from a religious tradition to create new meanings about religion and the religious person's duty to the natural world? What may be legitimately changed and what is considered sacrosanct? Under what situation or settings can a REMO reinterpret a tradition so that it now applies to the environment? To what extent does activists' cultural and institutional embeddedness constrain, enable, or channel their innovative work? I draw on insights about the strategies and processes of cultural innovation from work in the sociology of religion, organizations, and social movements to develop answers to these questions.

Innovation and the Constraints of Institutional Embeddedness

Studies of cultural innovation have identified a number of different practices through which actors engage in the creative process and have staked out a contextual argument to explain when innovation is likely to occur. There are several common strategies by which actors create new meanings or forms of culture. One body of work looks at the practice of bricolage, in which actors "tinker, borrow, improvise, experiment, and recombine existing elements" to create new forms of gastronomy, new commercial products, or religious rituals and identities.[3] Roof's work on the new spirituality illustrates how actors "reconnect" or rediscover lost religious traditions, "reframe" religious truth, or engage in re-traditioning or "creat[e] new cultural traditions that provide alternative visions of spiritual and ethical life."[4] Social movement scholars often conceptualize innovation in terms of various forms of "framing." A frame is an interpretive schema that defines injustices, offers new solutions, and provides a rationale to act by combining extant, ideologically congruent but disconnected frames ("bridging"), or redefining existing values and frames to fit a specific context ("transformation").[5]

One limitation of this body of literature is that the criteria, constraints, or facilitators of innovation are not fully specified or conceptually developed, and innovators seem to be able to exercise great agency as they seek to develop new forms of culture or new institutions. In a recent essay on cultural revitalization and fabrication, Snow and his colleagues argue that "resources and schemas are potentially polysemic, and thus can be interpreted and used in different ways by different actors."[6] While this is a plausible theoretical claim, it ignores the very real possibility that new interpretations will not make sense to an audience, or offend the audience to such a degree that sanctions are applied to the innovators. In others words, actors likely face any number of constraints when they attempt to create new frames of meaning. When scholars attend to the constraints on framing, they tend to emphasize how changes in the larger social context, such as wars, natural catastrophes, or economic downturns, may encourage innovation insofar as the changes to the context make old meanings, values, or rules less applicable. Thus during "unsettled times,"[7] or periods when there are "disturbances in the moral order,"[8] actors will reach into their tool kit bricoleur-style to create new rules and new ways to make sense of changes in the social world. Yet context alone is not as powerful or predictable as the literature suggests. Religious environmentalism emerged throughout the course of about fifteen years with few instances when REMOs' cultural work and periods of rapid social change or anomie lined up. Although REMOs were influenced by the growing public concern with climate change, or local ecological problems, my interviewees did not view such events as evidence that the social world was especially "unsettled" and thus an especially propitious time to mobilize.

REMOs are embedded in particular religious organizational fields whose members share understandings of the goals, how goals should be pursued, and who can participate.[9] Their location in specific religious fields more directly shaped their cultural work than macroevents or shifts in the political system. REMO leaders must address three important constraints in order to develop a viable green ethic: (1) the legitimate interpretive strategies available for use, (2) a religious group's orientation toward social issues both in terms of the rank ordering of issues and the type of issue addressed, (3) organizational settings that encourage or discourage innovation. First, in terms of interpretive strategies, religious bodies that rely on a literal hermeneutic or proof-texting will face more challenges in developing an environmental ethic if there are few direct mandates or references to "saving" nature in sacred texts or religious teachings than will those activists who are embedded in religious traditions that rely on nonliteral, more principle-based interpretive strategies.[10] Second, REMOs may be able to more easily create a green ethic if they are connected to religious groups for whom the environment is a legitimate issue, or if these groups emphasize the systemic causes of social issues. They may struggle to win support for their green ethic if their religious audience believes social issues are rooted in individual causes or believe religion should primarily attend to personal morality concerns (for example, sexuality).[11] Finally, organizational settings that foster multiplex and encompassing ties may allow for the creation of boundary-spanning identities or meanings.[12] That is, religious organizations that encourage inclusive religious identities and foster social ties that cut across traditions (for example, interfaith and ecumenical groups) may make it easier for activists to borrow and adapt as they attempt to create new meanings about the environment.

In addition, action within a given organizational or institutional field is guided and constrained by "logics of appropriateness"—a set of rules "concerning who may use [cultural scripts, forms, frames, or practices] for what purposes, under what conditions"—that are located in specific institutional and organizational fields.[13] These rules of the game set limits on what can be borrowed and how foreign cultural elements are to be integrated within a religious tradition. REMO leaders run the risk of losing legitimacy if they apply ideas or practices from the secular environmental movement that run counter to the norms and theological commitments of their religious audiences. REMOs that are strongly embedded in a particular religious culture and institutional field (especially REMOs tied to more conservative religions or one tradition) may face more limitations on their ability to borrow and apply ideas from secular or other religious groups. One way to get around this dilemma is to make sure that the elements borrowed are consistent with or easily connected to the existing set of religious ideas or follow the operative logic of appropriateness.[14]

Several institutional constraints shape REMO leaders' abilities to create a religious environmental ethic or tradition. These include the institutionally specific rules that govern cultural borrowing and adaptation, relationships of power and authority within a religious community, and the degree to which a REMO is strongly connected to a larger religious community or tradition. The content of a religious tradition or what is available in the religious tool kit may also constrain innovation. Religious traditions that have extant teachings about the environment, interpretive strategies that allow an activist to apply old meanings to new issues or settings, or a tradition that is flexible (that is, the tradition is understood as a work in progress and one that can be changed, elements added, or eliminated) will face fewer limitations on the innovative process than those that have little or no explicit teachings about the environment, that primarily rely on a literalist orientation toward sacred texts, and that see the tradition as being rigid or closed to alternations.

In the rest of the chapter I analyze the three main innovation strategies REMOs use to create new green religious traditions. I show how REMOs' different levels of embeddedness in specific religious fields affected their ability to craft a new eco-religious ethic and integrate environmentalism into existing religious traditions. More specifically, I show how REMOs' embeddedness within religious authority structures, audiences, and systems of religious meaning channeled and constrained the ways in which they framed religious environmentalism and attempted to build support, legitimacy, and resources for the new movement. I end the chapter with a discussion of REMOs' efforts to augment their new green ethic with a set of new ritual traditions tied to the sacred calendar and worship practices.

Mining and Reframing the Tradition

Although few religions have a deeply rooted environmental ethic, some REMOs are able to find pieces of their tradition that explicitly address nature or can easily be applied to environmental issues. This common strategy of innovation is less about creating something new than about rediscovering old cultural ideas and practices in the religious tradition and integrating them into a new green religious ethic. Becker describes this strategy as "mining the tradition," in which actors trawl through their religious tradition in search of elements within it that will "solve a particular problem."[15] In practice REMOs may simply aim to "reconnect" or draw their audiences' attention to some ecologically oriented text, principle, or practice that has been ignored in the tradition.

REMOS that have the strongest ties to a single tradition or religious body are the most likely to mine the tradition. The Jewish REMOs were particularly adept at using this practice, in part because some of their main religious laws focus on the sacredness of food and the ethics of food production and distribution. Every Jewish leader interviewed spoke about kosher, or kashruth, laws. In an essay from the Sacred Food Project, Rabbi Arthur Waskow, director of the Shalom Center, first explains that treating some foods as sacred, as well as the processes of producing and cooking food, has been a central way in which the Jewish people connect to God. Then he notes how contemporary ecological problems are raising concerns about the need to revitalize the rules of kashruth and encourage non-Jewish communities to follow them.

> Questions about the use of pesticides and of genomic engineering; of the burning of fossil fuels to transport food across the planet, meanwhile disturbing the whole climatic context in which the foods are grown; the misuse of topsoil and the use of long-term poisonous fertilizers . . . all these have raised profound new questions. . . . Should the category "kosher" be reserved for food alone, for what we literally put in our mouths and gullets? Or would it make sense in our generation to apply the basic concept to other products of the earth, other forms of "eating"—consuming?[16]

Waskow is referring to the notion of "eco-kashruth," a concept developed by the founder of the Jewish Renewal Movement. According to Debra Kolodny, who directed the Sacred Foods Project, "the word 'kosher' means fit for use, so if you get a cucumber out of the earth, but the ground was drenched with pesticides and herbicides, and the workers were exploited, and agribusiness bought up and decimated the livelihoods of 70 farmers, is that kosher?" In both cases, Jewish REMO leaders provide their constituents with a clear diagnosis of the agricultural problems and then suggest the tradition's answer—kashruth—is important but inadequate to address them. Instead the tradition needs to be expanded to fulfill the spirit of the rules about kosher foods, maintain the sacredness of food and its production, and connect to the more general orientation toward justice within Judaism. Hazon, a pan-denominational Jewish REMO, is the most active advocate of eco-kashruth. Its Community Supported Agriculture (CSA) program, called "Tuv Ha'Aretz," is a biblical play on words that means both "good *for* the land" and "the best *of* the land" (Ezra 9:12), was started in 2004 and by 2010 had expanded to forty sites across the United States. In 2005, it launched a food blog called "The Jew and the Carrot" and published a 130-page compendium of sources (texts) about Judaism and food. Hazon has two educational food programs and sponsors an annual Jewish food conference On Hazon's website, organizers link their concern for "eco-kashrut" to include the Sabbath (*Shabbat*), Sabbath year (*shmitta*), and year of Jubilee (*yovel*).[17] These Jewish concepts of rest historically required farmers to periodically let their land lie fallow, allow the poor to glean the fields, and restore freedom to the indentured (mainly agricultural workers). By doing these things Jewish communities would make the world a more just and healed place.

Nigel Savage, founder and executive director of Hazon, has written extensively about the Jewish tradition and environmentalism. He claims that his REMO turned its attention to food "because we think addressing these challenges [i.e., those articulated by Waskow and Kolodny] helps renew the Jewish community, and because we care seriously about applying Jewish tradition and the resources of the organized Jewish community to making a better world for all."[18] Savage aims to help Jews "reconnect" to ancient traditions. He believes doing so will not only persuade them to become more environmentally conscious, but it will also renew Judaism and the world itself.

Hazon's website and Savage's writings reveal a REMO steeped in Jewish tradition. However, nature or environmental issues are not explicitly named or the central focus of most of these elements from the tradition. Instead, Savage must "reframe" or reinterpret elements from his tradition to show how they are salient and applicable to the environment.[19] In an essay titled "Treasures within the Toolbox," Savage identifies several resources within the Jewish tradition that he claims speak to the environment and if applied will mobilize the community to act. In addition to the concepts of the Sabbath rest and kashruth, he discusses *brachot* (radical amazement), *lo tashchit* (needless destruction or consumption), and *tisha b'av* (Jewish responses to destruction). Savage begins the discussion of each concept by explaining the source of each (for example, Torah, Talmud, Mishnah) and thus signals their religious pedigree. After describing the meaning of each, he explains how the concepts apply to the environment. For example, he notes that practices of radical amazement, or *brachot*, are crucial for a new Jewish environmental ethic because "we cannot wish to protect and preserve something if we do not love it and see it in its full beauty and value. *Brachot* helps us to be radially amazed by the world; in so doing they are a key first step in environmental awareness and action." *Lo tashchit*, he notes, arises in a text from Deuteronomy about the destruction of fruit trees, but "the rabbis of the Talmud and their successors progressively expanded on this concept, so that "*lo tashchit*" — "don't destroy"—became a *halachic* [the body of Jewish laws] injunction. Like other parts of the Jewish tradition, *lo tashchit* is not absolute." The injunctions of *lo tashchit*, Savage notes, are unknown among most Jews today but may "become an important framework for renewed contemporary discussion about consumption and destruction" because they can compel Jews to see that their lifestyle and consumption choices should be based on religious prohibitions and exceptions rather than in custom or contemporary cultural norms.[20]

Savage reintroduces his Jewish audience to a lost tradition and legitimates its application to contemporary environmental problems by showing how the practice of reframing is a standard and acceptable part of more general rabbinic practices. Rabbi Jeff Soltar of the Shalom Center concurs with Savage, noting that "Jewish history for the past, oh, at least the past 2,000 years, what we would call rabbinic Judaism as opposed to Biblical Judaism, the rabbis were shameless in taking pieces of the tradition and applying them to build an approach to contemporary issues." Thus the logic of appropriateness or rules that govern the use of the Jewish tradition appears to allow for the extension and expansion of ancient traditions so they can speak to current issues.

More generally, Jewish REMOs rely on a justice frame to mobilize support, participation, and gain legitimacy from their constituents. Nearly every REMO leader uses the concept of "*tikkun olam*" — "healing, mending, repairing the world, improving society"—as his or her organization's central ethical principle.[21] COEJL's founding director, Mark Jacobs, notes that his group uses the concept as its "overarching framework" not only because it "has become fairly universal in the American Jewish community, but also because it captures the sense of Jewish responsibility for the people on the planet." Rabbi Nina Beth Cardin, leader of the Baltimore Jewish Environmental Network (BJEN), also claims that this is the one principle about justice that unites the wide range of Jews that make up BJEN, and it provides the means to see how saving the environment is intimately connected to tackling the injustices surrounding health care, globalization, capitalism, and poverty. In short, Jewish REMO leaders draw on both familiar and lesser-known parts of the larger Jewish religious tradition and demonstrate how these elements provide the rationale and mandate to yoke Jewish identity and lifestyle practices with environmentalism.

Bridging within Traditions

REMO activists also use a strategy akin to "frame bridging" ("the linkage of two or more ideologically congruent but structurally unconnected frames") to weave environmentalism into existing religious ethics.[22] The goal is to help religious groups and individuals see how more general religious values, concepts, and concerns (for example, charity) are connected or speak to environmental issues. The underlying logic of this strategy is to confront religious audiences with the commonalities between their traditional religious ethics and environmental ethics, and thus compel them to adopt or incorporate the environment into their ethical framework. The strategy relies on an "if/then" logic—"if you care about religious justice for the poor, for example, then you should care about global warming because the poor are more adversely affected by climate change, and we are commanded by God to take care of the poor." Thus bridging is different from reframing insofar as it involves less work altering the meaning of religious ideas and more work drawing connections between those ideas and environmentalism.

REMOs with limited proenvironmental resources to mine within their tradition often rely on this strategy. As with reframing, REMO leaders take great care to connect their environmental claims to central religious values, images, stories, and texts. REMOs that work within religions that are wary of environmentalism and/or have a tradition that largely ignores or minimizes nature must focus their creative work on bridging this gap of mistrust and indifference. REMOs whose constituency are Catholic or conservative Christians, in particular, heavily favor bridging within the tradition.

The two Catholic REMOs in the study work hard to explain how existing Catholic social teachings, especially from recent popes and the US Conference of Catholic Bishops (USCCB), should be applied to climate change. The Catholic Climate Covenant (CCC), a coalition of thirteen national Catholic Church agencies under the authority of the USCCB, organizes its work and explains its purposes around four church-wide principles: prudence, poverty, the common good, and solidarity. An extensive part of the REMO's website is devoted to catholic teachings that define these principles, identify how they have been articulated by important church leaders, and indicate how they apply to the environment. In a founding document of the CCC, the 2001 bishops' statement, "Climate Change: A Plea for Dialogue, Prudence, and the Common Good," the CCC's audience learns how the core ethical principles of Catholicism speak to the environment:

We especially want to focus on the needs of the poor, the weak, and the vulnerable in a debate often dominated by more powerful interests. Inaction and inadequate or misguided responses to climate change will likely place even greater burdens on already desperately poor peoples. Action to mitigate global climate change must be built upon a foundation of social and economic justice that does not put the poor at greater risk. . . . Working for the common good requires us to promote the flourishing of all human life and all of God's creation. In a special way, the common good requires solidarity with the poor who are often without the resources to face many problems, including the potential impacts of climate change.[23]

By anchoring its frame in the concern for the poorest and most vulnerable populations, the CCC connects its appeal to the rich Catholic social justice traditions that encompass Dorothy Day's Catholic worker movement, liberation theology's notion of the "preferential option for the poor," and the 1984 USCCB statement on economic justice and the US economy.[24] Doing so sends a signal to Catholic churches, agencies, and lay members that climate change is a legitimate concern and simply an extension of the church's ongoing work to alleviate poverty. During our interview, Dan Misleh, CCC's executive director, reiterates this strong, almost natural connection between concern for the poor and concern for the environment:

It's not just the environment for environment's sake. We're not just about saving the spotted owl; we're also about saving people, and people are the ones that are primarily impacted by environment. I mean other forms of life are as well, but if we believe that we are created in God's image and likeness and the pinnacle of creation, then our primary concern ought to be how people are impacted. . . . While we have lots of debates on climate change, about what it's doing to the polar bears or what it's going to do to amphibians in the rain forest, that's a very important discussion. But our added value, as a faith community, is what's it going to do for poor people or to poor people?

The CCC's effort to create a Catholic environmental ethic is remarkably holistic and integrative. In its overview of Catholic social teaching and climate change, the CCC reminds its audience of "our duty to cultivate and care for the gift of creation. . . . If we harm the atmosphere, we dishonor our Creator and the gift of creation." Then it provides a four-page set of excerpts from papal and episcopal documents that link stewardship with other Catholic moral teachings. One of the first excerpts is from Pope Benedict XVI's 2010 address to members of the diplomatic corps in which he tightly links the dominant Catholic concern with abortion and the culture of life with environmentalism: "How can we separate, or even set at odds, the protection of the environment and the protections of human life, including the life of the unborn?" The CCC's strategy appears to push the claim that Catholic environmentalism is simply part and parcel of an overall Catholic way of engaging with the social world and congruent with the tradition's long-standing focus on alleviating human suffering. This way of reframing Catholic social ethics may allow the CCC to expand its pool of supporters beyond the usual liberal peace and justice types to include rank-and-file Catholics more interested in acts of charity or preventing abortion.[25]

Like their Jewish counterparts, the Catholic REMOs are deeply embedded in their tradition, and they marshal it to legitimize environmental activism. However, they are not nearly as free to reinterpret, modify, or apply the ethical teachings as are Jewish REMOs, which is one reason why they are more likely to adopt bridging rather than reframing. As noted above, the CCC was founded by the USCCB and serves as the agency to promulgate the bishops' messages about climate change and other ecological matters. Misleh uses a political metaphor to describe the relationship:

> The basic division of labor has been, and will continue to be, that the Bishops' conference sets the public policy agenda, and the Catholic Coalition on Climate Change help[s] promulgate that public policy. . . . I implement what they say. I help when the Bishops put out a particular piece of, or at least get some direction to how they want the public policy to unfold. Then with my connections with diocese and other organizations across the country, I can help them generate the ground troops, I guess you'd say.[26]

In her study of green sisters Taylor recounts her conversation with a staff member of the USCCB that reveals the ways in which Catholic REMOs' embeddedness in the institutional church limits how they can frame and develop religious environmentalism. According to this staffer, the official church position on the environment is Christ focused and human focused, and often the work of "Green Catholics" moves too far away from these twin foci. The Roman Catholic hierarchy is suspicious of the emergent ecotheology of Thomas Berry and other green Catholics because it endorses biocentrism, which places nature over and above humans. The problem in their view is that it may confuse Catholics about the true source of revelation about God and lead to idolatry: "It became easy to identify the movement [Catholic environmentalism] with naturalism (the doctrine that all religious truths are derived from nature and natural causes and not from revelation. . . . [which] is not in accord with Catholic teaching."[27] Similarly, Warner describes the official Catholic environmental ethic as "a tempered form of anthropocentrism" insofar as traditional Catholic concerns about respect for human life and care for the poor drive church initiatives on the environment.[28] When I asked senior staff member Tim Kantza if the NCRLC works with groups that promote the kind of eco-theology and spirituality of Thomas Berry or some of the women's religious orders, he quickly identified the theological limits that guide and restrict such partnerships:

> Some would contend that some of his proponents are a little pantheistic. I don't see that. I think there's a thin line there that you've got to be careful about. I think basically Thomas Berry has opened a new discussion, and it's a new way of viewing the universe and our connection with it, that is appealing but not against church teachings. . . . The religious women are the green wing of the church, and they're really moving us forward toward sustainability. There are some that have stepped over the line. . . . What gets inter-mixed in [their approach] is eco-feminism, which tends to get people into some trouble too because some of them stress that as opposed to the other, and then it ends up being the role of the woman in the Catholic Church, and that just gets into a big mess. And then everybody's misunderstood. So we try not to weave all those different connections together.

While both the Jewish and Catholic REMOs are strongly embedded in their respective traditions, the two follow different logics of appropriateness. As I noted earlier, the Jewish tradition allows for and encourages the on-going reframing of tradition, and the lack of a centralized authority system that can direct or veto their work gives the Jewish REMOs significant cultural freedom. Conversely, the powerful official ecclesial system of the Catholic Church more significantly limits how the Catholic REMOs operate. Two fundamental rules guide their efforts to use their tradition. First, the bishops do the heavy lifting—they identify the traditions to be used and how they are to be interpreted and applied to environmentalism, and the REMOs articulate and implement the bishops' vision. Second, green Catholicism must be grounded in the larger Catholic ethic around life that encompasses the poor, the marginalized, and the unborn.

Evangelical REMOs also are significantly constrained by the logic of appropriateness, authority structures, and theological commitments of their communities. Since the 1980s, the Christian right has focused its social justice concerns on personal morality issues and, in particular, on the "trifecta" of "pornography, abortion, and same-sex marriage," while remaining silent on the environment.[29] Wilkinson notes that evangelical environmentalists are trying to broaden the scope of evangelical social ethics so that "climate care is one among a number of connected social issues."[30] This finding dovetails with other recent scholarship that suggests that evangelicals, especially younger age cohorts, are adopting more liberal attitudes toward a variety of personal morality issues and becoming increasingly concerned with larger collective issues, and the environment in particular.[31] Several evangelical interviewees discuss the generational divide within the broader evangelical community and note how they seek to mobilize twenty- and thirty-something evangelicals. A Rocha, USA, Restoring Eden, and the Southern Baptist Climate Initiative actively target young adults. Jonathan Merritt (SBCI) claims that his generation "is more pro-life than the generation before us. The difference is that our definition of what pro-life means is much broader. We don't believe that being pro-life simply means opposing the culture of death. It means supporting life wherever we find it. I am pro-environmental because I am pro-life, and these are life issues." Similarly, Restoring Eden's (RE) Peter Ilyn notes that his audience is coming to support creation care "because this is an extension of being pro-life. And being pro-life to them means not rushing to war; it means worrying about mercury in the bloodstream; it means knowing that an African American kid dying of asthma on a school playground is [a] pro-life [issue]."

Another bridging strategy Ilyn and his REMO use is to help their evangelicals connect contemporary environmental activism with early childhood experiences of nature. Ilyn argues that he needs to help them understand that caring for creation—an ideal that is revived in those childhood memories—is a fundamentally Christian act and one that is anchored in their core values and theological commitments. He recalls his own experiences operating a booth at Christian rock 'n' roll festivals in which young adults stop by Restoring Eden's booth and voice their relief that someone in the evangelical world shares their concerns about nature. RE's attempt to bridge young evangelicals' love of nature and the core values of their faith is evident in the following interview excerpt:

> [i.e., young proenvironment evangelicals] told [me] stories of being out in nature. And I realize that what they have in common is that they love nature. And that's when I realized we've got to make that safe because these spiritual epiphanies that happen in the wild are a historic part of Christianity. . . . So in a sense three concepts came back: to love, serve, and protect. And in our kind of phrasing, you know, make your heart bigger, your hands dirtier, and your

voice stronger. So we reinforce that kind of thing. . . . And that's where I think the right-to-life conversation becomes valid again as well because we show them how this is consistent with other core values they hold. And that's the right-to-life argument, and even the kind of justice or Evangelism arguments are not paradigm shifts, they're paradigm expansions. So we're asking our audience, look, these are your core values, and if you expand them to include nature, then they're consistent. And if you shrink them so that it doesn't include nature, then not only is it illogical, it creates cognitive dissonance.

Ilyn refuses to define evangelical morality in a way that privileges personal morality concerns over collective issues but instead aims to show how RE is simply incorporating environmentalism into the evangelicalism's culture of life. Other evangelical REMO leaders also are aware that they must show how and why environmentalism is consistent with their decades-long pro-life commitment. Rusty Pritchard says that the Evangelical Environmental Network (EEN) seeks to help evangelicals "make the connection between being pro-life and fighting poverty, being concerned about justice for the poor and climate change." The EEN's environmental health campaign explicitly references the pro-life connection with its title, "Mercury and the Unborn Child," and its rationale for the program links the pro-life argument, Jesus's Great Commandments, and environmentalism: "Christians are called to protect life, and for us that includes the unborn. Jesus taught us to love our neighbors and treat others as we would want to be treated. Protecting the unborn and children from mercury poisoning and air pollution is in keeping with Jesus' commands. It is time to stop the mercury poisoning of the unborn."[32]

The effort to frame the call to care for creation as biblical is at the core of every evangelical REMO's cultural work. The SBCI's Jonathan Merritt invokes Jesus's commandments in order to communicate the religious imperative to act against climate change, but he also refers to a classic Sunday school adage and other references such as biblical passages to reiterate how environmentalism is part and parcel of the Christian life:

Jesus said that the greatest commandment is to love the Lord your God and second is to love your neighbor. And so I think anytime you're about to make a decision, down to the smallest decision to purchase a light bulb, you say what puts God first; what puts others second, and what places me in a distant, distant third. The unfortunate thing is, and when you reverse that, when you place me first and God last, that's what we call sin. I think a lot of Evangelicals, especially in public policy positions, are living a life of sin because you ask them why they're making a decision, and they begin to talk about your pocketbook; they begin to talk about economic wealth; they talk about materialism; they talk about things that run absolutely counter to the teachings and the ministry of Jesus Christ, placing themselves first. How do I take care myself; how do I put money back into my pocket, and then others come in a distant second. God doesn't even really come into the equation, except to drop in verses out of context to support our own greed. I find that when you teach people that little mechanism there, and as kids we learned J-O-Y—Jesus first, others second, yourself last. When we live a life of JOY, where we put God first, what will make his creation flourish? What will make it better? What will make me a good steward in the sense of Genesis 2:15 [the passage in which God charges Adam and Eve to "till and keep" the Garden of Eden]?

Merritt's comment also points to the ways in which evangelical REMOs anticipate the possible lines of resistance their targeted audiences may exercise. He appeals to the Christocentrism of evangelicalism by showing that protecting the environment is a way to follow Jesus's two commandments (to love God and neighbor from Mark 12:28–31), and thus he makes it difficult for detractors to make the case that environmentalism will pull evangelicals away from their fundamental religious duty. Then he addresses the counterargument that environmentalism is not biblically sanctioned by linking it with God's command to be good stewards of creation from the book of Genesis. Finally, he acknowledges evangelicalism's "anthropology of sin" but turns the tables by arguing that the kind of politics endorsed by the Moral Majority (his allusion to evangelicals' support of conservative economic policies that place profit over love of neighbor) rather than environmentalism leads to sin.[33]

All of the evangelical REMO leaders I interviewed are very aware of the internal resistance they face. In their effort to create a new green evangelical ethic, they address criticisms that environmentalism is nature worship and pulls evangelicals away from more biblically important activities such as evangelism and stopping abortion. The EEN's FAQ web page starts by first laying out the notion that the environment "is not separate from or subordinate to God's creation, and that Jesus' death and resurrection not only offers salvation to humans but reconciles all of creation to God."[34] Evangelical REMOs must show how central theological values and biblical commandments serve as a bridge to environmentalism. Ultimately they offer their audience a way to fulfill their existing ethical duties of stewardship and service to God by acting to limit climate change that most strongly affects the world's poor.[35]

Yet evangelical REMOS do not simply aim to link environmentalism to core values and theological principles. They supplement this kind of substantive bridging with a process-oriented strategy. In their account of the production and reception of the National Evangelical Association's 2004 statement on the environment, Djupe and Gwiasda show how NAE leaders made environmentalism plausible and legitimate by adhering to the proper process through which evangelicals are supposed to come to policy decisions. They write:

> Evangelicals emphasize a distinct process for reaching decisions—a process that involves intense, personal reflection, both on the Bible and with God through prayer. Only through this process do evangelicals trust that a decision, either their own or one made by others, is valid. Signaling that the decision was reached via prayer or scriptural reference provides a cue to members that a message provider is a legitimate in-group member.[36]

Interviews with evangelical REMO activists are peppered with biblical references and quotations, and their websites extensively quote scripture to justify their brand of Christian environmentalism. For example, EEN's FAQ page provides numerous quotations from the Bible to legitimize its central claim that environmental action is integral to the Christian life. In the fall 2004 newsletter of the EEN (*Creation Care*), several participants at the 2004 Sandy Cove conference describe the canonical process they followed as they developed the new evangelical ethic of creation care: "We have meditated on Scripture"; "We prayed, talked together, and listened to some of the world's leading Christian naturalists and scientists"; "It was striking to see God's Spirit moving among us."[37] By invoking scripture, prayer, and the work of the Holy Spirit, evangelical activists support their substantive claims for Christian environmentalism by demonstrating that they follow the correct procedures that lead them to this novel position.

Pritchard and Merritt also discuss the important role that credentialing plays within the evangelical community and how they aim to generate credibility among their audiences by listing the evangelical luminaries who support creation care. Wilkinson notes that this kind of "grass tops" strategy is a crucial way of signaling legitimacy to the larger evangelical world given the lack of a "formal institutional hierarchy" like that of the Catholic Church that can give Christian environmentalism an official imprimatur.[38] As Rusty Pritchard notes, "The strategy is to reach influential leaders who can then be leaders in their own community." The hope is that these leaders will take the message of creation care to their own churches, colleges, and parachurch groups and mobilize rank-and-file evangelicals. Strong embeddedness in evangelical religious values and theological claims, in evangelical hermeneutics and decision-making processes, and in a skeptical audience profoundly shape how evangelicals craft their new ethic of creation care.

Several liberal and mainline Protestant REMOs also report how their embeddedness in both tradition and the official church structure shapes their particular green religious ethic and efforts at frame bridging. The Unitarian Universalist Ministry for Earth (UUMFE) draws its inspiration from the "Seventh Principle" of the Unitarian Universalist Association ("respect for the interdependent web of all existence"), the denomination's roots in the naturalism of the American transcendentalist tradition, and its historic commitments to pursuing social justice. The UUMFE's leader, Rev. Katherine Jesch, argues that these commitments push the REMO to frame religious environmentalism around justice rather than stewardship:

> We tend not to use the language of stewardship very much because most of that Christian theology implies humans as outside and above nature, and having this kind of paternal responsibility. That is not our theology. Our theology is that we are fully integrated; we are a part of nature. And our Seventh Principle says the interdependent web of all existence, of which we are a part.

Framing its appeals for support in terms of stewardship seems unimaginable to Jesch, as it would fundamentally violate the theological commitments of her denomination and thus discredit the work of the UUMFE among church members. This concern to adhere to the UUA's principles and not adopt the more common and thoroughly Christian notion of stewardship is important since the denomination has moved to the heterodox side of Christianity, and its members may not even consider themselves Christian.[39]

Leaders of Quaker Earthcare Witness (QEW) exercise a similar level of caution as they advocate for a distinctly Quaker approach to environmentalism. Elaine Emmi (one of the founders of Quaker ecological activism) identifies both the theological ideals (reverence for creation, working for peace and justice) and principles that guide church politics (working via consensus and deliberation) in her response to my question to describe the Quaker approach to environmentalism.

> Quakers don't have set doctrine, but they do things through a lot of committees. We ask ourselves a lot of questions. We highlight issues by asking questions; they're called queries. And often we'll have discussions about it, or there might be actions that result from that. But the point is—sometimes

there are no right answers. Everybody has their [sic] own ideas about what works for them. So that's one of the reasons you don't try to have set doctrine, but create a dialogue. . . . We work to integrate into the beliefs and practices of the religious society of friends, the truth that God's creation is to be respected, protected and felt in reverence for its own right, and that the truth that human aspirations for peace and justice depend on restoring the earth's ecological integrity.

Although there is widespread agreement around the central importance of peace and justice for members of the Society of Friends, if QEW were to violate the process of producing a testimony on the environment they would lose support of their core audience. Hollister Knowlton, former board chair of QEW, concurs with Emmi, noting that the REMO must follow denominational norms, especially those that protect its radical congregational polity, grassroots orientation, and emphasis on the centrality of individual conscience. QEW's leaders recognize that they must adhere to these norms to be seen as legitimate and important by members of the denomination, even if it means specific ecological policies or practices are not implemented.

These examples also illustrate the ways in which REMOs' bridging efforts are constrained by their embeddedness in specific audiences. REMOs' abilities to creatively modify and expand their traditions are constrained by the responses and/or anticipated responses of their audiences. Renee Rico, executive director of Presbyterians for Restoring Creation (PRC), commented on the vital role audiences play in shaping how she goes about her work communicating Presbyterian USA's official teachings and positions on the environment. Given the theological heterogeneity of the denomination, she relies on a justice frame that stresses the need to ameliorate the negative effects of climate change on the poor among liberal or progressive church audiences, and stewardship among more evangelical Presbyterians:

So if I'm working with a more Evangelical group, I talk about being a steward of creation; that language just communicates better. They understand what you're saying. It's not trying to pussyfoot about it, but it's simply saying here's a frame I know you already understand, and how do I communicate. People who are already doing lots of justice work and are really involved in peace-making; the eco-justice piece is a frame that connects to their frame.

She continues by noting that it is crucial for the mission and success of the PRC that the REMO "connect foundationally with what people believe and connect with the things [that is, agriculture, transportation, food] people care about." This powerful concern about audiences' receptivity to a new green religious ethic also informs the work of the REMOs who rely on other strategies of innovation.

Bridging across Traditions

Many ecumenical and interfaith REMOs struggle to find a message that will resonate with audiences from several different traditions. Rather than drilling down into a particular tradition and then showing how the tradition provides the requisite scriptural or theological warrants for environmentalism, many REMOs emphasize the shared general values of different religions in their new green ethic. Most commonly, interfaith REMOs talk about at least one of four common ideals or values shared by the three Abrahamic religions: the sacredness of creation; stewardship; justice for the poor or most adversely affected by environmental problems and intergenerational equity. Bea Moorhead, executive director of Texas Interfaith Power and Light (TIPL), summarizes this practice in her response to my question about how she speaks to different kinds of religious groups:

> I would say working across all the religious traditions, there are four main reasons that people of faith come back to environmental concerns. One of them is the idea of Genesis, the idea of stewardship; whether you're a Christian or not, you probably hark back to that idea of stewardship. One of them is the least of these; the idea of the impact on other people of environmental action and degradation, things like that. One of them is inter-generational—equity, which is a broadly held religious concern. . . . And then there is this idea of the holiness of all creation, which is expressed in different ways in different faiths but always is at least sort of suggesting to people that quite apart from anything that has to do with you, we have reason from our religious traditions to think that the creation is important to the Creator just because of itself, its own inherent merit. We look at Job as probably a really good and convenient example of that in the Abrahamic tradition. But the Muslim tradition certainly has a very strong thread of that.[40]

Moorhead identifies the common ground constituents of TIPL share and argues that these provide the reasons for cooperation and joint action. In her essay for the Sacred Food Project, Clare Butterfield (Faith in Place [FIP]) stresses how the Abrahamic traditions share a common anthropology that must compel Jews, Christians, and Muslims to take on the role of "caretaker for the earth—of steward for the fertility of the earth," and then connects this to the three religions' shared teachings about justice for those who are hurt by oppressive, dangerous, and unhealthy agricultural practices and outcomes (for example, the farm workers and consumers made ill by pesticides).[41] Butterfield provides scriptural warrants for environmentalism and does so in order to signal an interfaith audience that FIP is aware and respectful of each tradition, and that the shared ideals in their sacred texts should bind them together to promote sustainable agriculture.

A handful of interfaith and ecumenical REMOs touch on the sacredness of creation in their appeals for support and action but then emphasize the environmental issues facing the world. Here the argument is that the scope and seriousness of global environmental problems, especially climate change, transcend religious differences. Cassandra Carmichael, who leads the National Council of Churches Eco-Justice Group, illustrates this approach. She notes first how a shared belief in the sacredness of creation helps the NCC's constituent bodies support the REMOs' programs, and then emphasizes the urgent nature of the current ecological crisis to intensify the call to care for creation:

> I think the awesomeness, the amazingness, the joyfulness of creation, and the mystery of creation sort of helps people transcend their little minor idiosyncrasies. . . . I think it's when you interact with creation and you're interacting with God in creation, it transcends—it sort of moves you out of your particular situation. . . . And we all have to have food, we all have to have water, we all have to have shelter, so there's a unity there in knowing that we are all at the end, we're all dependent upon the inter-relationship of creation. And lately the environmental situation has been seen as urgent, and there's a need to address it. The environmental crisis creates an urgency that demands that people leave the pettiness or the particular arguments aside. . . . They [i.e., various ecological crises] move people beyond their immediate selfish needs. And they have to work together if they're going to get it done.

The Genesis Covenant, an interfaith REMO whose mission is to mobilize the entire religious community in the United States to reduce its greenhouse gas emissions by 50 percent before 2020, articulates the issue's appeal in more simple terms: "We have to put aside our religious differences to focus on what we share in common: *an understanding that global climate change is threatening our world and commitment to do something about it*" (italics original). This alternative mobilization strategy eschews a deep grounding in specific traditions and seeks to use the environmental context as the bridge between different religious groups.

Although the large group of interfaith and ecumenical REMOs may not be as deeply tied to a particular religious tradition as the Jewish, Catholic, or Evangelical REMOS, they are equally attuned to their audiences. These REMOs face a particularly difficult challenge as they seek to gain supporters and mobilize action because they have no institutional constituency. No one is a member of an ecumenical or interfaith church, and thus it is not always clear what religious values, texts, or traditions will pull in different religious individuals or groups. Too strongly emphasizing one tradition over others may alienate potential supporters, while ignoring specific religious values may make it difficult for potential supporters to understand why they should join or support the REMO. Recent research on faith-based organizing finds that interfaith organizations commonly "seek to transcend religious differences by focusing on shared values and common goals."[42] The strategy of emphasizing common interreligious beliefs and the need to fix pressing environmental problems that threaten all religious groups appears to follow this practice.

Interfaith and ecumenical REMOs thus try to tailor their message to particular audiences in order to impress on them that the REMO's mission and goals are aligned with the extant core values, theological principles, or social justice priorities of a given religious body. As suggested by the public speaking strategy of the PRC, some REMO leaders note how they emphasize a justice message with more liberal Protestant groups and stewardship or creation-care message with evangelical and more conservative Protestants. For example, Bea Moorhead of TIPL reports that she rarely talks about eco-spirituality because it is not a welcome perspective among her Bible-belt audiences: "It's generally tough for [them] to get their heads around the idea of maybe the trees and rocks are important to God." Moorhead also notes that she does try to tailor her message to highlight the most important values of a particular faith community: "If I am talking to a Jewish community, I would talk about, for instance, inter-generational equity because it is particularly important in the Jewish tradition. So I kind of bring to bear the things that are most important in particular churches."

The bridging strategy of the ecumenical and interfaith REMOs aims to encourage members of each tradition to discover how to be more fully Jewish or Christian or Buddhist while also working toward a shared religious vision for a greener earth. As Cassandra Carmichael notes: "The folks that didn't necessarily see themselves as tree-huggers or environmentalists are starting to see how, from a place of faith, environment is important. . . . We're sort of in the middle of that shift. And I think that'll continue on, so that folks see you can't be fully Christian and you can't be fully Jewish unless you also recognize your stewardship and justice call to God's creation." REMOs act as the bridge that moves individuals and religious bodies from a place in which environmentalism is not important to a place in which it is part and parcel of the religious life. REMOs' keen awareness of their audiences' religious and environmental beliefs and their ability to make their message resonate with each audience or to find a new message that will transcend sectarian perspectives is a crucial part of an emergent movement's success. However, "Audiences affect what social movement groups do (and vice versa)," and may exercise a sort of veto power by granting or withholding legitimacy from cultural innovators.[43] Sometimes REMOs' audiences exercise this kind of veto power directly, but often the REMOs anticipate possible objections and limit their creativity. Crossley argues that "although constraints, by definition, bite whether or not the actor recognizes them as such, exactly how this plays out is dependent upon whether the actor recognizes them in advance. Anticipating negative sanctions and acting so as to avoid them has a very different consequence, for example, to failing to do so and stumbling into a conflict situation."[44] Successful REMOs are more adept at anticipating objections to their innovative work and minimize resistance by exercising caution. If the innovation too radically departs from sacred texts and the tradition, or if audience members cannot see how the new green ethic is related to or is born from existing religious teachings, texts, or religious practices, they will be less likely to accept it as legitimate. Therefore most REMOs work hard to show how environmentalism is not a side issue but core to the identity, ethical mandates, and practices of a particular religion.

Which type of bridging strategy REMOs use depends on the nature of their embeddedness in specific religious traditions, authority structure, and audiences. Their different embeddedness also compels REMOs to adhere to different logics of appropriateness as they attempt to develop their new green ethics. Evangelical REMOs that bridge within their own tradition are careful to preserve core values, teachings, and practices generally shared within the evangelical world and try to explain how and why the core of the tradition should be extended to environmentalism. This is most evident in their focus on the biblical teachings about stewardship, evangelism, and sin, but also in the way in which they publicly follow the legitimate biblicist interpretive strategy and decision-making process in order to secure the endorsement of key leaders within the religious world. Doing so signals to their audience that the REMO's call to integrate environmentalism into the life of faith is congruent with an evangelical worldview and way of being. In addition, the insular nature of evangelicalism means that REMOs within this religious world are completely dependent on other evangelicals for support, resources, and participation. If they are unable to mobilize other evangelicals, then these REMOs are out of business since they will not be able to secure support from nonevangelical audiences. Conversely, interfaith and ecumenical REMOs aim to mobilize support from a wider range of religious groups and thus are not as constrained by specific religious cultures. However, their reliance on multiple religions compels them to adopt the common ground strategy and to avoid offending any particular audience by excluding or misinterpreting their particular traditions.

Bricolage and the Creation of Eco-Rituals

Bricolage is the least commonly used innovation strategy, in which actors "recombine already available and legitimate concepts, scripts, models, and other cultural artifacts that they find around them in their institutional environment."[45] It is a riskier strategy than bridging, reframing, or mining. The innovation (for example, the new religious ecological ethic or ritual) may fail to resonate with the REMO's audience because it is seen as incongruent with or in violation of the existing tradition, or because it simply is not meaningful to the targeted audience. The use of bricolage poses considerable risks if the REMO aims to integrate faith with ideas from science or deep ecology, if it tries to blend Western and non-Western religions, or if it tinkers too radically with existing worship rituals. Reworking worship or ritual life is especially problematic because it threatens to interfere with the ways in which the believer and the sacred interact and thus makes religious boundaries and identities less certain.[46] REMOs that are strongly tied to a particular religious community may face discipline or the lack of official support if their efforts at bricolage are deemed too radical. For example, Bender and Cadge describe how Catholic nuns borrow the meditation techniques of Buddhism but not the accompanying religious philosophy because to do so would make them theologically and doctrinally suspect in the eyes of other nuns and the official church hierarchy.[47] REMOs rarely use the ideologies of eco-spirituality or deep ecology because both advance biocentric claims—that nature possesses intrinsic value, that all species are of equal value, and that humans are not separate from and superior to nature—that challenge the anthropocentrism of many religions.[48] Thus certain types of bricolage appear to violate fundamental religious beliefs about God and the appropriate relationship between God and humans, which seems to curtail this practice among many REMOs.

Unsurprisingly, REMOs that represent multiple traditions are more likely to use bricolage. They do so carefully and in ways designed to avoid alienating their diverse audiences. They often suggest modest changes to existing religious narratives or rituals. For example, Voices for Earth Justice (VEJ), an interfaith REMO, draws on Thomas Berry's "New Story" about the emergence and unfolding of the cosmos, but does so in a way that is sensitive to the different audiences it hopes to reach. VEJ has strong ties to the Catholic Church in Michigan and many supporters who are Catholic, and while Patricia Gillis (VEJ) is aware that Berry's "New Story" is not a canonical part of Catholic theology, she believes it is an important addition to the religious environmentalist's tool kit. It pushes people to reconsider the nature of reality, because the "Universe Story" claims nature and the cosmos itself are sacred, and removes humans from the center of the story. It pushes her audiences to reconsider the dominant creation story and to adopt a more eco-centric view.

> I think we need to add the universe story to our repertoire. . . . I have various versions of it—I have a kids' version; I have a version that's very scientific; I have a version that's more watered down, and more mystical, for nuns or people like that who are into contemplation. . . . One year, for our retreat "Connecting with Creation 4," we did the universe story. And we had a large rope on the floor, and we had big candles at various points along the access, and we had narrators, and we had music in the background, and we had people sit on the periphery . . . and someone walked it. It's very powerful because the rope is like 460 feet long, and humans are in the last inch. It gives you a sense of perspective of the creation, of the universe, planet earth, and the evolution of humans. We're the newest kids on the block . . . and it is a real good way to diminish anthropocentrism.

Gillis tailors the story to different audiences to maximize the story's chance of resonating with each. While she doesn't dismiss or replace the older stories explicitly, she seems to suggest that this new cosmology needs to be integrated into the old cosmologies from all of the Abrahamic religions that VEJ serves.[49]

Gillis's account also illustrates how REMOs are creating new types of green religious rituals. Several REMOs blend climate science into Jewish or Christian holiday rites and, in doing so, infuse religious ritual with political meaning. For example, the Shalom Center's Green Menorah Covenant turns Hanukkah, an eight-day commemoration of the rededication of the Holy Temple in Jerusalem, into a rite that raises Jews' awareness of climate change and encourages them to reduce their use of oil and coal during the holiday.

Rabbi Jeff Soltar explains that the program is presented in ways that anchors covenant is in Jewish history, sacred texts, and theology, and thereby assures his Jewish audiences that reworking the tradition can help Jews reclaim their heritage and live more full Jewish lives.

> The menorah is one of those recognizable Jewish symbols and green means environment, so in two words it captures the eco-Jewish roots. . . . The symbol of the Hanukkah menorah as an instance of one day's worth of oil, lasting for eight days, is an example of doing more with less that can inspire us. . . . I mean one of the aspects of Hanukkah is that it's about energy use and taking bold action in the space of limited oil. . . . So another basis for the green menorah is each arm of the menorah is a different path of action. For each arm there are two different elements—one is the need for personal or institutional action, which would be what an individual could do or their synagogue could do whether it's carbon reduction, alternative fuels, transportation, land use, and the other is for public advocacy [on these issues].

The campaign uses the image of an olive tree in the shape of a lit menorah and a quote from the book of Malachi about Elijah's call to lead the Jewish people to turn their hearts to the Lord (YHWH). Failure to do this will result in the Lord's "utter destruction" of the earth. It also includes a brief statement about the threat of utter destruction posed by global warming and urges Jews to make a new covenant to end climate change.[50] This illustrates the Shalom Center's careful use of elements from the Jewish tradition to expand the significance of the ritual and add a new layer of practices without radically altering it. Those observing the holiday could still light the candles, recite the blessings, and sing or pray according to the customary practices, but they add prayers to address climate change and add new activities during the holiday to reduce global warming. Similarly Georgia IPL offers its members study guides to rework the Advent or Hanukkah rituals by changing one incandescent light bulb to a compact fluorescent bulb each time a candle is lit. Its guides include scriptural passages, eco-themed prayers, and short discussions of climate change and generally attempt to explain why individuals of faith should take action.[51] Again, by adding new meaning to extant religious practices without altering the fundamental meaning of the rite, these bricoleurs offer a safe way to incorporate environmentalism into worship.

REMOs may adopt this kind of tempered bricolage that adds but does not replace or blend multiple religious ideas to avoid resistance that syncretism can engender. An alternative approach that also is sensitive to the interests and sensibilities of its audiences is "do-it-your-self" bricolage. Here REMOs provide extensive liturgical, scriptural, and theological resources for their diverse audiences and encourage them to select prayers, texts, or ritual acts from an interfaith smorgasbord that will be the most meaningful and consistent with their existing beliefs and practices. For example, the GreenFaith "Green Worship Planning Tool" provides its members with a template for environmentally focused worship and a host of resources by specific faith tradition as well as a forty-five-page set of interfaith prayers.[52] This strategy then puts the responsibility for blending elements from different religious traditions on the congregations and individuals who attend a REMO green worship workshop to download the materials and shields the REMO from charges that they are unduly changing or damaging a particular religion's traditions. In short bricolage is constrained not only by the degree REMOs are embedded in specific audiences and how much creative license audiences will grant, but also by the nature of the tradition out of which they work. Bricoluers, as Altglas argues, start from a given religious tradition or traditions and are constrained "by the meaning and content of resources that are being used."[53] Thus the tradition shapes the new green rituals by providing the raw materials and the parameters within which innovation must fit—the new Hanukah rituals around the green menorah are familiar even while adding a new layer of meaning to the old ritual and symbol.

Conclusion

In this chapter I have described how REMOs create a new religious culture of environmentalism by using several strategies of innovation. I have also shown how their work to create a new green ethic and religious eco-traditions is differentially constrained by the ways in which they are embedded in particular textual, theological, and ritual traditions; religious authority systems; and audiences. REMOs face different combinations of constraints that push them to use some strategies but not others. For example, REMOs that are embedded in strong authority systems, that are steeped in and committed to preserving a specific tradition, and that are trying to mobilize an audience that also closely adheres to the tradition may be more likely to engage in bridging or reframing (like the Catholic REMOs). The case of evangelical REMOs illustrates how the content of the tradition and the rules about using it push them toward the bridging strategy. First, their tradition has few explicit connections to environmentalism and instead emphasizes evangelism, charity for the poor, and personal morality/culture of life concerns, and thus these REMOs focus on showing how environmentalism is consistent with and even a way to realize these other religious goals. Moreover, they must rely on the interpretive practices and decision-making rules (part of the evangelical culture) that govern evangelicals in order to gain credibility and legitimacy with a skeptical audience. Finally, evangelicals are very dependent on the support of their audience since attempts to recruit nonevangelical supporters will damage their standing within evangelicalism. Thus bricolage is not a viable option as most evangelicals would consider the resulting religious ethic to be unchristian, and reframing runs the risk of too radically changing the tradition.[54]

Interfaith and ecumenical REMOs also rely on bridging, but largely because they are loosely embedded in multiple religious traditions and thus face fewer pressures to conform. In many ways they must be religious "omnivores" who possess a sound working knowledge of multiple traditions in order for them to bridge or blend them. They routinely identify the shared religious principles and values that can unite a diverse audience. They also are embedded weakly in religious authority systems, since they are formally independent of national religious bodies and have greater freedom to experiment, borrow, and blend various traditions into their green ethic. At the same time, their audience tends to be far more open to non-Western religions and spirituality and more broadly conceived than Catholic, Evangelical, or mainline Protestant audiences, which encourage bricolage rather than mining and reframing.

Jewish REMOs are tied to a rich religious tradition that not only provides a wide array of resources from which to construct a new Jewish environmentalism, but also a tradition that allows for and encourages constant tinkering and change. In addition, Jewish REMOs are embedded in an audience that may not be fully aware of the ties between nature and their tradition and in a weak authority system that is not empowered to limit the REMOs' creative work.[55]

In short, REMOs are pushed toward specific innovation strategies based on the degree to which their tradition(s) provide resources about the religious foundations of environmentalism and the rules that dictate how REMOs may extend or modify the theological and ethical resources of a given tradition. For example, a REMO anchored to a tradition that relies on biblical literalism likely will be compelled to rely on bridging since the Bible rarely speaks directly about the environment, and such a hermeneutic severely limits on reinterpreting sacred texts. Innovation is not only shaped by the availability, rhetorical force, or resonance of cultural elements that are borrowed and reworked, but also by the influence of specific audiences. REMOs' efforts to create green religious cultures hinge on audiences' threat of discrediting or withholding legitimation if REMOs fail to abide by the institutional rules for borrowing, blending, and reframing religious culture. In the next chapter, I show how these new religious eco-cultures and REMOs' institutional embeddedness encourage and limit interorganizational coalition building.

FIVE

Coalition Building and the Politics of Cooperation in the Emergent Movement

As the director of the National Council of Churches' Eco-Justice Ministry, Cassandra Carmichael is connected to a wide variety of religious and non-religious groups that are part of the larger environmental movement field. Her work of organizing and overseeing a host of programs on ecological problems, convening meetings with various stakeholders, and disbursing grants to REMOs gives her a unique perspective on the nature of interorganizational relationships within the new movement. The NCC was founded in 1950 to promote Christian unity and develop a shared voice and collaborative projects on pressing social issues of the day. The eco-justice working group started in 1983. Since then, it has produced a library of faith-based resources for denominations and parachurch groups and engaged in advocacy on everything from toxic waste to climate change. Currently the eco-justice program works with nearly seventeen member denominations, twenty REMOs, and several state councils of churches.

Carmichael began our conversation about the NCC's role in the movement field by describing the two types of religious organizations with which she works and highlighting her role as a broker[1] of information and resources:

There's [sic] two types of organizations we work with. There's the ecumenical agencies, like the Washington Council of Churches or the Oregon Ecumenical Council, the Maine Council of Churches, Minnesota Council of Churches. . . . They're affiliated with us in a way. They have a representative on our governing board, so we definitely work with them as much as possible. That might just mean that I have interactions with them at a meeting; it doesn't necessarily mean that we give them funding. With the para-church groups [i.e., REMOs], it's the same thing, but the relationship is more informal. And there aren't that many para-church organizations. There's just a handful at the larger state level, and we work with as many of those as we can. And when we're doing state-specific work or projects, then I try, as best I can because I know they struggle for money, to float some money their way.

Her comment suggests a few possible reasons why REMOs join coalitions: to secure resources, to implement a particular environmental project or participate in an NCC campaign (such as the 2005 interfaith global climate change campaign), and to fulfill obligations that arise due to existing interorganizational relationships. In addition to providing seed money to REMOs, she often provides "strategic" advice to new or smaller REMOs. Such information may mean simply directing organizations to NCC eco-justice resources, but it may also require her to take a more direct role by explaining how to organize interfaith dialogue and build new alliances, or helping them develop their own educational campaigns. When I asked her how she manages such a diverse set of relationships (formal and informal, denominational and parachurch) and roles (broker, banker, project manager, consultant, convener), she described the operating principles and interorganizational dynamics that make the NCC's partnerships and coalitions successful:

The overall mission of the program is to do a couple things: to resource congregations and people of faith so they can work on eco-justice issues, and then the other function is that the program serves to help form an ecumenical table for the different members of the NCC to work together. And they've been working together for over 25 years actually. There are some people that are representatives on the working group that have been there since the beginning. . . . The NCC has historic black churches, mainline Protestants, and Orthodox communities . . . and they don't always agree on some of those simple things like what music do we play in church, and how do we take communion. But it has been a blessing to see them come together around this issue of eco-justice, and to see how they have come to value each other's space, walk, and tradition in a way that I don't think they would have if they hadn't been working on this issue together for so long. And there's a deep respect; it's not the Episcopals trying to make the AMEs Episcopal or the Presbyterians trying to convert the Orthodox into their way of thinking. It's actually a very genuine respect for each other's faith traditions. . . . And also becoming educated about each other's traditions; each of the traditions brings a gift to the table. It's sort of Biblical. There's a passage in the Bible that talks about each person is given a gift, and perhaps each of the different denominations or communions is also given a gift—a way to care for creation, and the eco-justice thing in a unique and powerful way. . . . So I feel that the working group has done a really good

job of exemplifying what I believe to be true of ecumenism, which is not trying to create one consensus church, but in the coming together of fellowship
and work and mission, that they have manifested through the body of Christ,
and each part of the body is distinctive and important.

The NCC's partnerships seem to work because they are based on mutual
respect and trust among members, while also granting autonomy to individual organizations. It appears that the Eco-Justice Working Group manages to minimize the risk of losing each REMO's specific identity and religious commitments, which in turn lessens the chance that each constituency
will withdraw its support.[2]

However, this same kind of trust and respect has been more difficult to
build with nonreligious partners. Carmichael speaks to the "collar problem"
that the NCC and many REMOs experience when they explore partnering
with secular environmental movement organizations (SEMOs). Too often,
she explains, some secular group calls up looking for the NCC to provide
a minister (that is, someone wearing a clerical shirt that has a detachable
collar that is part of the uniform among Catholic priests and some Protestant ministers) to attend a rally or a letter-signing event in order to claim
that it has the support of American religions for its particular cause. When
this occurs REMO leaders often feel that they are being used and their own
religious voice will be silenced by their secular counterparts. They are wary
of joining partnerships with them. However, REMO-SEMO relations have
improved in recent years, in part due to the ways in which coreligionists have
set boundaries and refused to play by the SEMOs' rules, and in part because
secular movement organizations have begun to see the unique benefits
REMOs bring to the larger environmental movement. Carmichael continues:

> I think they've gotten much better about, in my opinion, about being respect-
> ful, but maybe that's because I've had a little chit-chat with them. . . . And I
> think Paul [Gorman of the NRPE] has also been really good about represent-
> ing the religious community to the environmental community, and setting
> boundaries for them and letting them know "Hey, you know what, we can't be
> bought. You can't just use us; you can't just roll over us." . . . The faith-based en-
> vironmental voice has reminded the secular environmental voice that there's
> a moral dimension to their argument. The secular environmental community
> has a gift to offer, and they should be doing that gift. And they shouldn't try
> to become religious or spiritual or, I mean not that they're amoral, but that's
> not their call. That's not what they're good at; they're really good at advocating
> and dealing with the science and providing knowledge about environmental

issues. . . . If everyone sticks to what their goal is, and we work together, I think that's the strength of it [i.e., REMO-SEMO partnerships].

Carmichael's comments suggest that organizations in the emergent movement are still struggling to establish and maintain boundaries that differentiate themselves from the secular side of the larger movement. These boundary concerns reflect the challenges REMOs face in building credibility with religious audiences who may be unsure that environmentalism is a legitimate religious issue. She also suggests that religious and secular movement organizations may have different goals, cultures, and norms about organizing that make it difficult to forge working relationships with one another. Yet her account of interreligious and inter-REMO cooperation is incomplete; she does not discuss the difficulties she faces working with Evangelical groups and why they are reluctant to work with mainline, liberal, and Catholic REMOs. In addition, her account paints a rosy picture of inter-REMO cooperation that belies the more common story about REMOs' reservations about working with organizations that don't share similar values and theological views, the tenuous nature of cooperative relationships among REMOs, and their relative isolation from one another.

More generally her account raises key questions about how religious environmentalism, as a whole, is organized and operates. With which groups will REMOs cooperate, and what criteria lead them to establish or eschew alliances? How do concerns about resources, ideologies, and constituencies affect interorganizational cooperation? Which REMOs serve as brokers or bridges that connect themselves to one another, and what is the nature of the ties that are formed? How does the structure of the religious environmental field shape the kinds of interorganizational relationships among REMOs?

In this chapter I describe the interorganizational networks among REMOs, and between REMOs and secular environmental movement organizations (SEMOs), and examine how different types and degrees of embeddedness influence cooperative relations. I look at the ways in which REMOs' embeddedness in religious authority systems, audiences, and cultures makes some REMOs unwilling or hesitant to enter into alliances while encouraging other REMOs to seek out new partnerships. More specifically I show how the same kind of constraints that shape REMOs' cultural work, discussed in the preceding chapter, also influence their cooperative relationships, while the imperatives of niche activism, described in chapter 2, push some REMOs to enter alliances in order to gain much-needed resources. Then I describe the relationship between religious and secular movement organizations and show how REMOs limit such alliances in order to manage their boundaries, maintain the integrity of the mission and identity, and/or avoid alienating their audiences who consider such alliances religiously problematic. I finish the chapter with a brief discussion about the limits of cooperation within religious environmentalism and across the larger movement field.

The Structure of the Religious Environmental Movement Field

Movement coalitions vary widely in terms of duration and degree of organizational formality, size and complexity, range of activities, and efficacy. Most simply, a movement coalition or alliance involves two or more social movement organizations (SMOs) that "work together on a common task."[3] REMOs commonly are involved in short-term, often informal, alliances with one another in order to share resources and information and, less commonly, to work on a specific campaign together. REMOs have stronger and more lasting ties to one another and other religious organizations than they do with SEMOs, and having an information-exchange relationship often encourages REMOs to engage in protest together because the prior relationship has built the requisite trust between partners.[4]

The most common type of tie between REMOs involves the exchange of information. In practice this often means that REMOs share best organizational practices regarding mobilization techniques, funding sources, potential partners (both religious and secular), educational programs and curricula, and discuss the possibility of joint action. These relationships tend to be informal and short-lived. Figure 5.1 portrays the complete information exchange network within the movement. The figure identifies three important features of the movement field. First, like many social movement fields, the REMO network has a core-periphery structure.[5] The thirteen largest nodes, constituting the core, consist of four interfaith and five ecumenical agencies, one Catholic, one Jewish, and two mainline Protestant organizations. These thirteen REMOS have an average of fourteen ties to other REMOS, while those on the periphery average just 2.4 information-exchange relationships. An additional interfaith REMO, the NRPE, lies just outside the core, but this position misrepresents the significance and role this REMO has played in the history of the movement as one of its key founders and brokers. The core, including the NRPE, represents 62 percent of all information-exchange relationships (192 of 310 ties).

The core is divided into three subgroups. Four REMOs (NCC, COEJL, NRPE, RP) connect REMOs to one another, serve as patrons who provide funding to smaller or resource-poor REMOs, and/or act as organizers who design specific programs or actions and then recruit other REMOs to

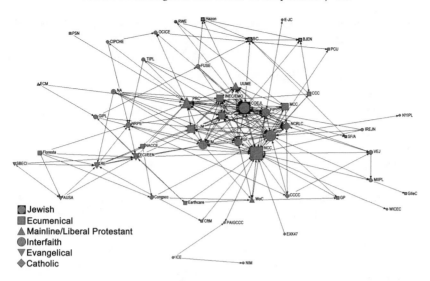

Fig. 5.1. Information Exchange Network of the Religious Environmental Movement (Source: Stephen Ellingson, Vernon Woodley, and Anthony Paid, "The Structure of Religious Environmentalism: Movement Organizations, Interorganizational Networks, and Collective Action," *Journal for the Scientific Study of Religion* 51 [2012]: 266–85. © 2012 The Society for the Scientific Study of Religion.)

participate.[6] Rusty Pritchard from the Evangelical Environmental Network aptly describes the EEN's role as broker in this way: "We do some programs on our own, but a lot of what we would like to try to be is the place where folks can hear and get connected to the environmental work, care for creation, and all of its partner organizations." The NCC's Eco-Justice Program operates in a more expansive broker/coordinator/organizer role within the emergent movement. For example, the NCC created religious climate change and antitoxin programs and then recruited state-level REMOs or helped found them (for example, the Wisconsin and Pennsylvania Global Interfaith Climate Change Initiatives) to implement the program. The NCC provided the grant money that paid for staff and programming for each of the state organizations. They also organize a wide variety of conferences that bring together REMOs, representatives of national religious bodies, and secular movement organizations around such issues as climate change, water and air pollution, and toxic waste. Religious leaders meet one another at these events, exchange ideas, and may even be encouraged to create new movement organizations (for example, the Chesapeake Covenant Congregations was formed after several of its founders attended the NCC's 2005 water stewardship training conference). Finally, the eco-justice program produces materials for green worship events across the Christian calendar and curricula on such topics as green cleaning, caring for local watersheds, and environmental racism. These materials have been used by thousands of congregations and individuals who may obtain them through another REMO.[7]

The second set of REMOs within the core is connected as members of a new coalition that meets regularly to exchange ideas and develop a shared frame and action agenda (EJM, EM, INEC, GF, NCP, MCC, COEJL, RP, NCC, UUMFE, PRC, and NCRLC). Peter Sawtell, one of the conveners, notes that the key goal of the meetings is to become "aware of what niches are out there and who has strengths in what areas; that have the staff and publishing resources, for example, to put out top quality resources." He hopes that the meetings will build "a sense of community and trust" that will allow them to develop a common eco-justice frame upon which to enter into joint activities. Thus relationships based on the sharing of information help REMOs identify and solidify their own brand of niche activism and use the resources of their peers to augment their repertoires of mobilization and contention rather than expend their own scarce resources duplicating the work of their peers. Finally, a few REMOs are heavily engaged in a number of alliances and partnerships that involve advocacy at the national level (NCRLC, NCC, NRPE) or on multiple ecological issues (FIP, INEC). For example, the NRPE identifies key policy issues and legislation on which REMOs will speak and then helps its members (which include the NCC, COEJL, and the EEN) develop the talking points, educational campaigns, and mobilization tactics to be used within each member's community.

Second, outside of the core, there are a handful of cliques based on shared religious affiliation. The evangelical REMOs are all connected to one another on the far left side of figure 5.1 with very few ties to nonevangelical REMOs. The Jewish REMOs are clustered together on the far right side of the figure, and all are connected to COEJL and a few other groups that either have strong ties to American Judaism (FUSE, Sacred Foods) or to REMOs that work in the same geographical area (for example, BJEN is tied to the Chesapeake Covenant Congregations, an ecumenical REMO). Although a shared religious identity seems to be important for some REMOs, overall shared religious affiliation does not predict information-exchange relationships.[8] Even the most sectarian REMOs have information-exchange ties that cross religious boundaries, which suggests that REMOs are willing to engage in dialogue around common environmental concerns in order to learn from one another and assess the possibility of taking joint action.

Third, most of the REMOs on the periphery have few ties and usually are connected only to one of the core REMOs or to a small number of

REMOs with which they share a common ideology about the environment or work on a common issue in the same geographic area. Nearly 20 percent of REMOs do not have ties to other REMOs. The intensely local or regional focus of most REMOs (recall that more than 70 percent work at the local, state, or regional levels) means that they may not be aware of other REMOs or they may believe a REMO far removed from their location cannot provide much help. For example, Katy Hinman of Georgia Interfaith Power and Light notes the difficulty her organization has working with other state IPLs and the parent REMO, the Regeneration Project, because her southern location creates a more challenging context in which to mobilize religious support and action: "Here in Georgia, and this is something that we kind of have to remind those folks in California, global warming is not accepted as fact. . . . It's kind of challenging working [with the Regeneration Project], since the national office is located in California. I think they sometimes forget that Georgia is not really on the same level as California." Hinman's comments suggest that context may profoundly shape alliance formation, and that context may override shared identities and ideologies held by REMOs. This may be especially true when a REMO's audience has a dissimilar understanding of environmental problems and how best to address them.

Figure 5.2 depicts the joint action network. Although the core-periphery structure persists, it is a less complex and dense network than the information exchange network. Ties are sparser, REMOs appear to be more closely allied with others that share the same religious affiliation or commitments, and more REMOs work alone. The core also is smaller and anchored by a few large, national organizations. Some REMOs join together in classic social movement activities such as participating in marches or rallies, or engaging in lobbying and litigation. However, the more common forms of collaboration focus on educational endeavors, such as offering climate change courses for congregations, or offering a fellow REMO help developing programs to promote energy efficiency within religious organizations. As I discuss below, the joint action network is attenuated largely because these types of relationships pose greater risks to REMOs, especially those whose legitimacy and survival depend on maintaining support of an audience that is difficult to mobilize.

Explaining Cooperation or the Lack Thereof in the Religious Environmental Movement

Why is the religious environmental field organized around a small core? Why are most REMOs weakly connected to one another, while a few are very

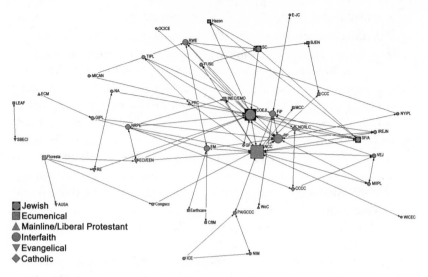

Fig. 5.2. Joint Action Network of the Religious Environmental Movement (Source: Stephen Ellingson, Vernon Woodley, and Anthony Paid, "The Structure of Religious Environmentalism: Movement Organizations, Interorganizational Networks, and Collective Action," *Journal for the Scientific Study of Religion* 51 [2012]: 266–85. © 2012 The Society for the Scientific Study of Religion.)

active building or participating in alliances? Why are ecumenical and interfaith REMOs far more likely to engage in collaboration than REMOs from a single tradition or from conservative Protestantism? Why are evangelical REMOs willing to talk to other REMOs but do not collaborate on joint actions with many other REMOs, even other evangelical groups? Building coalitions is a challenging, but often necessary, endeavor for social movement organizations. Entering a coalition opens up a SMO to several risks, including the loss of autonomy, the loss or credibility among key constituencies, and the need to set aside its own particular goals to realize those of the coalition.[9] SMOs, then, must weigh the risks and benefits of cooperation with those of remaining independent and will be encouraged to join a coalition under a number of conditions. The literature identifies four primary factors that encourage or discourage alliance formation: culture, resources, political opportunities and threats, and preexisting ties between activists or organizations. While it is important to consider how these broad factors influence why REMOs enter into cooperative relations, they do not tell the entire story. REMOs do not always operate in ways that the theories predict. They tend to mobilize around values rather than issues, which can make some alliances unpalatable. In addition, since they are less focused on the political arena than most SMOs, they may not be attuned to threats and opportunities for collaboration. Finally, REMOs' embeddedness in particular traditions, authority systems, and audiences has a profound effect on their willingness to build or enter partnerships with other REMOs.

Culture, Networks, and the Influence of Audiences

Scholars of movement coalitions agree that SMOs must share some common goals, a commitment to a shared ideology, and/or overlapping identities, to consider cooperating with one another.[10] This may be very important for a REMO because it is embedded in a set of ongoing relationships with other religious groups (for example, denominations or congregations) who may exercise authority over the REMO. Even if a REMO has only informal ties to a national body, it may feel obliged to adhere to the theological beliefs and standards of the religious body, especially if doing so confers greater legitimacy on the REMO among elites and rank-and-file members of the community.[11] REMOs, then, may seek to appear legitimate to their religious constituencies by maintaining their theological and ideological distinctiveness, and thus avoid collaborating with REMOs that do not share the same affiliation or theological frame about the environment.[12] However, shared commitments to justice and norms about inclusivity may promote interorganizational ties between REMOs. Brown and Brown observe that groups from mainline Christianity often follow a practical theology that emphasizes justice, which, in turn, encourages them to engage in interreligious dialogue and cooperative activities.[13] Similarly, Lichterman finds that groups from liberal theological traditions may possess the vocabulary and group-building customs that encourage ties that bridge religious groups. Such customs may include a willingness to engage in "social reflexivity" (a willingness to engage in discussion about the relationship between religious group and society), a commitment to drawing inclusive group boundaries, muting direct religious talk, and using instrumentally oriented language.[14]

It is not surprising that ecumenical and interfaith REMOs are more likely to be involved in cooperative relationships—both information exchange and joint action—because they embrace this liberal ideal of inclusivity; because their organizational form is based on partnerships; and because they build their green ethic out of common values, ideas, and practices from various religions. In my interviews, these REMO leaders repeatedly claim that widely shared ideals about stewardship, the sacredness of creation, or justice make it possible to work with one another. Andy Burt, the director of the eco-justice programs for the Maine Council of Churches, notes that the MCC's ties to other REMOs are built on "the deep stream of justice, reverence, and mindfulness that is part of every religion. We have to find the common stream and common words that bind us." The central role that shared culture plays in uniting REMOs is also evident among the Jewish groups. Each group grounds its work in several key values, laws, or principles of Judaism—*tikkun olam*, lo tashchit (do not destroy or waste), kashruth laws—and its leaders note that this shared framework makes close relationships and joint action possible.

In addition to shared theologies of the environment, REMOs who cooperate with one another may also share a common set of rules about how to organize and work together. The executive director of the Interreligious Eco-Justice Network (IREJN) notes how both shared theological views and a shared style of relating or engaging in coalition politics allows this interfaith group to function:

> The people that are coming to the table are used to being at the table. They—
> Jews, Methodists, UUs, UCC—they're all aware that there's multiple claims to
> truth. They're all aware that there's some things that unite them. . . . They're
> not going to sit there and go, "I'm not listening to the Methodists." And to
> some extent, they're already in some limited coalitions and collaborations
> with other faith groups.

Some scholars show how sharing the same expectations about how to engage in the practical activities of movement politics are integral to alliance building.[15] This is especially important when SMOs face strong barriers to cooperation due to conflicting ideologies or goals.[16] For example, the NRPE created a new set of rules its members could follow that allowed for differences in values and ideologies. Gorman, the NRPE's executive director, notes that the organizing rules anchored around the principle of "walking together separately."[17] Instead of finding common theological approaches to environmentalism, his organization allows each member organization to address the same issue in ways that fit within the strictures of its own tradition. Gorman formulated this principle during a meeting that brought Catholic, mainline and Evangelical Protestant, and Jewish leaders together for the first time. After hours of conversation, he realized that the theological differences might be too great to overcome if national religious bodies were really going to get involved, but then it hit him that they were all committed to addressing climate change and a few other issues but could not agree how to do so together:

We basically identify some of the priorities that would be important for the religious community to explore. . . . And then we're just sort of saying that we're all going to do this separately and together, and this is the time of the meeting in which I scribble down two phrases—walk together, separately; and be ourselves, together—[these] phrases find their way into the bylaws of the partnership. But that's also the organizing gestalt.

The key organizing principles for the NRPE are to allow each member to retain its religious distinctiveness while working together to engage in advocacy at the national level, and to help groups (for example, congregations) within their tradition develop their own environmental programs. For example, in a joint program on mercury poisoning, the Evangelical Environmental Network is free to frame the issue in terms of mercury's threat to unborn children, which keys into the broader ideology about the culture of life that is important among evangelicals. At the same time, the NCC is able to frame the issue in terms of justice, as mercury poisoning more powerfully affects disadvantaged populations, a claim that resonates strongly with many of its liberal constituents. This type of strategy protects the core values of a REMO and its constituents, while also helping to organize and even celebrate religious differences.[18] Joint-action ties that threaten to compromise the core, such as cooperating with a REMO that constituents do not regard as holding the same values and beliefs, tend to be avoided, which may help explain why these types of relationships are far less common than information-exchange ties.

The relationships among Evangelical REMOs illustrate how shared values and ideologies are not always sufficient to engender interorganizational cooperation. The evangelical REMOs were not nearly as tightly connected to one another as the literature leads us to expect. Although they share similar commitments about scripture, ethics, and guiding religious values, Evangelical REMOs rarely participate in common protest campaigns, advocacy, or even educational events (and apart from the EEN, they rarely engage in joint action with any other REMOs at all). One reason for the weak relationships among Evangelical REMOs may stem from the intraevangelical differences about which social issues are legitimate concerns for evangelicals. Recall that personal morality issues (for example, sexuality or abortion) not only have been at the top of the list for American evangelicals since the 1980s, but for some, they are also the only social or ethical issues worth taking a public stand on. Ilyn and Merritt both acknowledge that this difference in goals is also a generational conflict; younger evangelicals are more likely to recognize the environment as a legitimate religious issue than older generations. In 2007, Peter Ilyn of Restoring Eden issued "An open letter in reference to the call of Dr. James Dobson and fellow family ministry leaders for the National Association of Evangelicals to either silence or dismiss NAE vice president Richard Cizik." In the letter Ilyn criticizes Dobson and a number of evangelical luminaries (such Tony Perkins of the Family Research Council; Gary Bauer, Campaign for Working Families; and Jerry Falwell) for "ridiculing and criticizing" creation-care advocates and for "reinforce[ing] a notorious perception in America that the evangelical community does not care about the world's environmental crises and the suffering and loss created by them." In the most strongly worded paragraph of the letter, Ilyn summarizes the differences between creation-care evangelicals and the body of evangelicals in the Dobson camp:

> We at Restoring Eden are astonished that the Dobson group would criticize Mr. Cizik's wider beliefs about what are "great moral issues" and "traditional values," and yet be blind to the extreme narrowness of their own list of legitimate concerns for evangelical Christians. We know of no Scriptural support for narrowing the list of Christian moral issues to abortion, the integrity of marriage, and the teaching of sexual morality to our children. Further, for the Dobson letter signatories to fail to recognize that care of creation is a vital aspect of valuing and saving human life is a strong indication that these leaders are ignorant of the meaning and scope of humankind's stewardship role regarding the Lord's creation—and of evangelical Christians' egregious failure to address creation's degradation. It is as though they have so insulated themselves from the other biblically mandated moral issues that they have made spiritual blinders for themselves.[19]

Ilyn's critique highlights the conflicting goals between the older and younger generations and illustrates his effort to expand and reframe evangelical ethics to include environmental activism. REMOs like Floresta and the EEN, which aim to recruit older evangelicals or work within the old guard, may have trouble working with a group like Restoring Eden because their audiences use different criteria to justify environmentalism. As Ilyn notes: "What the EEN and the older evangelicals are trying to do is kind of add an amendment to the bottom of the Evangelical thing that says, 'the environment matters.' Where what we're trying to do is, in a sense, rewrite a whole different contract." In other words, Restoring Eden does not simply wish to add the environment to the bottom of a long list of issues evangelicals will aim to ameliorate, but to make environmentalism central to evangelical faith.

Doing so signals RE's commitment to evangelicalism's overarching commitment to life and fidelity to God's rule as laid out in the Bible.

Ilyn's comment also highlights a second barrier to greater cooperation among evangelical REMOs. EEN, SBCI, and Floresta seek to build legitimacy and resources by recruiting elite support, while Restoring Eden and A Rocha USA favor a grassroots approach. Thus different styles of building community, organizational norms, or different understandings of what constitutes "good politics" may make it difficult for evangelical REMOs to move beyond information-exchange relationships.[20] Intraevangelical differences about the nature and goals of activism around public morality may make it difficult for them to forge working relationships. Disagreements over the appropriate focus of reform (personal morality or structural/systemic problems) and legitimate process for achieving their goals (changing individuals' hearts one at a time versus challenging unjust systems of power) create a wariness toward intraevangelical cooperation.[21]

Evangelical REMOs also are less likely to forge ties, especially joint-action ties with REMOs who represent nonevangelical communities.[22] Strong theological differences are a significant barrier to cooperation, along with concerns to protect in/out group boundaries and incompatible cultures of activism as suggested above.[23] Leaders of the evangelical REMOs recognize that working with organizations that do not share the same beliefs will hurt their standing and support in the evangelical world. For example, Rusty Pritchard, a staff member of the Evangelical Environmental Network (EEN), states: "Evangelicals have a much harder time cooperating with groups like the NCC where the feeling among Evangelicals is that theology just doesn't matter to those folks. . . . [They] deny that Jesus was born of a virgin and deny that he was bodily resurrected, yet would still claim the name Christian. And that worries Evangelicals, and so they find it hard to cooperate." Similarly, Jonathan Merritt notes that threat of losing his audience's support severely limits organizations with which he can partner. In particular, liberal and mainline Protestants are off the list of acceptable allies as are ecumenical and interfaith organizations:

> If I were to work with the National Council of Churches, I'd be hard-pressed
> to find any Southern Baptists who will come with me. We're a very cloistered,
> cliquish group, which is why no one was able to do what I did because it had
> to come from within. [Those] outside of Evangelicalism have no credibility.
> They're immediately written off as somebody who doesn't get us; somebody
> who doesn't hold a value; somebody who doesn't believe in scripture, so it's
> unfortunate.

Many other REMOs also find that theological barriers or the lack of a shared culture limit the extent of their partnerships. And it is not simply that the REMOs themselves hold different religious values about the environment; their core audiences hold very clear views about which groups are legitimate partners. When a REMO violates the rules and moral categories its audience holds about the type of partners they view as appropriate for the movement organization, they often face a "legitimacy discount."[24] A number of leaders claim that they will lose credibility with their key constituencies if they enter into the wrong kind of partnerships, and thus feel pressed to preserve their theological distinctiveness and maintain religious boundaries. For example, informants from Catholic REMOs say that they need to exercise great care in working with groups that might support policies that conflict with the church's teaching on life. The policy director of the National Catholic Rural Life Conference notes that the "USCCB probably would not encourage churches to show *An Inconvenient Truth*, not so much about the content of the movie but more about its connections with Al Gore and his being pro-choice." The NCRLC's close ties and dependence on the Catholic hierarchy therefore set firm limits about interorganizational cooperation.

Leaders of ecumenical and interfaith REMOs also identify the ways in which dissimilar belief systems limit cooperation. For example, a leader of Earthcare, a southern mainline Protestant REMO, notes how the conservative religious climate and constituency compel them to "stay pretty middle of the road theologically" and "emphasize biblical warrants for creation, care, and stewardship." To do otherwise, or to partner with groups that might push for a more justice-centered approach, would run the risk of alienating supporters. For REMOs tied to conservative Protestantism or Catholicism, the public nature of joint action with theologically dissimilar groups poses significant risks and the potential loss of legitimacy and support among their core constituencies. This concern pushes them to avoid joint action altogether or to manage such relationships by seeking out significant operational autonomy, as in the case of the groups in the NRPE.

Another cultural factor that limits cross-REMO cooperation is the strong commitment most organizations have to fulfilling their narrowly defined religious missions. Recall that REMOs are primarily focused on raising awareness among individuals, congregations, and other parachurch organizations of the same tradition(s). The sparseness of REMOs' interorganizational ties is consistent with a focus on promoting a faith-based environmentalism within specific religious bodies. This internal focus is evident in the following remarks from the executive director of the Unitarian Universalist Ministry for Earth:

> We're not an environmental action group. We don't see ourselves that way and we don't project ourselves in the world that way. We're a ministry, and we are providing spiritual resources. . . . We're trying to help congregations turn their environmental values and activist leanings into productive work to make change.

Interviewees consistently echo the NRPE's primary goal "to weave environmental vision and values across the entire fabric of religious life, once and for all." Most REMOs, therefore, have few ties to one another because their missions focus on mobilizing a specific religious population in a particular place. Building alliances with other REMOs may not further their narrow religious goals. For example, Quaker Earthcare Witness aims to persuade Quaker meetings that environmentalism is congruent with the historic practices and beliefs of the Society of Friends and to influence the advocacy work of the Friends National Committee on Legislations. Given the parameters of their mission, there seems to be little incentive to seek out information or other resources from non-Quaker REMOs. REMOs who work on issues tied to specific places (for example, the Puget Sound watershed or forests of northern Virginia) may not seek out partners from other REMOs because these groups may not understand the ecology or organizing challenges they must address in the specific eco-space in which they operate.

Finally, REMOs struggle to forge alliances because they simply do not know one another or they are not connected to REMOs in the core of the network. A number of studies indicate that SMOs are more likely to form coalitions when leaders of the organizations have a history of working together, and therefore have built a significant level of trust and shared goals.[25] Several interviewees discuss how they are hamstrung in their efforts to work with particular types of REMOs (and the larger religious communities they represent) because they do not know any leaders in those REMOs. Reverend Fred Small, executive director of Religious Witness for the Earth (RWE), laments that he has not been able to bring Evangelical or Islamic organizations into his coalition because he doesn't know any activists from those traditions: "Unfortunately, it so often comes down to personal relationships. I don't fraternize with them otherwise, so I can't just pick up the phone and say, 'Hey Joe, hey Sally, I've got this great thing happening; do you want to come?' That would be my coming out of nowhere, with no relationship, and then there are the theological differences as well." Small's last comment underscores how REMO networks and culture are interrelated; he doesn't personally know Evangelical or Islamic activists in part because the deep theological differences between RWE (a liberal, interfaith organization, and Small is an ordained minister in the liberal Unitarian Universalist denomination) and leaders from these other two traditions. Ultimately, Small has no prior ties that he can activate, nor does he have ties that can link him with Evangelical or Islamic REMOs. Evangelical leaders also report that they are not well connected to activists or religious groups outside of the evangelical world.[26] Again, the insular nature of evangelicalism, coupled with the strong boundaries that separate conservative and liberal Protestantism, makes it difficult to develop working relationships. There seem to be few REMOs or REMO leaders that are positioned to connect evangelicals and nonevangelicals.

However, preexisting interorganizational and/or interpersonal ties among a small number of REMOs (largely from the core of the movement field) facilitate both information exchange and joint action. REMOs that share board members are significantly more likely to enter into information-exchange alliances than those who do not.[27] Leaders of the state IPLs whom I interviewed report that the annual meetings of the Regeneration Project have helped them build relationships with one another, and those relationships become the basis for working on projects together. For some REMOs, preexisting denominational ties lead to cooperation (for example, Web of Creation and Congregations Caring for Creation have strong ties to the Evangelical Lutheran Church in America), while other REMOs rely on their work in other movement coalitions (for example, the peace movement) to help them develop alliances. Sue Ellen Lowry, who runs the Noah Alliance, explains that the REMO emerged following routine conversations among religious activists: "Well, key members of the religious community, frankly leaders of groups in the religious community who care about conservation and talk to each other all the time, were talking in late 2004, and we realized that there was real Congressional threat to the Endangered Species Act. So we raised some funds to help create the Noah Alliance." Activists' personal and organizational networks appear to influence the level of cooperation within the new movement insofar as these ties reduce uncertainty about alliances, define certain partners as acceptable, and offer opportunities for collaboration. Lowry's comment also illustrates how REMOs enmeshed in diverse religious networks (that may include denominational bodies, state-level religious lobbying organizations and parachurch groups, along with other REMOs) activate those ties to capitalize on new opportunities or threats to their goals as the political situation changes.

Threats, Resources, and Pragmatic Politics

Much of the literature on coalition formation emphasizes how the broader political context, both opportunities and threats, can spur or prevent movement alliances. New legislation or public policy about the environment, shifts in party control of state or federal government, or economic downturns may be seen as threats to REMOs' goals and push them into alliances.[28] However, only a few REMO coalitions have been prompted directly by events in the broader political sphere. For example, the Noah Alliance formed to oppose congressional legislation to weaken the Endangered Species Act; RWE activated its partners in 2007 for a protest walk to raise awareness about the growing threat of climate change; a number of REMOs that work on climate change developed partnerships with the emergence of green energy companies during the 2000s. A small number of REMOs participate in the annual religious lobby day at various state legislatures but may do so independently or as part of a short-lived coalition, but such activities play a minor role in the overall agenda of the REMOs. Because REMOs focus on religious goals and their religious audiences, they are less attentive to and affected by affairs in the political sphere. This may be an important reason why their partnerships are not triggered by or focused on external factors as frequently as they are for secular SMOs.

REMOs are far more likely to seek out partnerships to address concerns about resources. Many operate with small budgets and staffs and are pressed to husband or increase resources. This motivates them to reach out to peers who may be able to supplement their stretched resources. The executive director of Presbyterians for Restoring Creation (PRC) describes how the search for resources and its focus on relatively narrow niches in the larger religious movement motivate cooperation, especially information-exchange relationships: "We are always sharing resources ecumenically because, you know, everybody's operating on very slim margins, so there's no point in duplicating what somebody else has already done." Many of the smaller REMOs contact one of the core REMOs (NCC, Earth Ministry, Eco-Justice Ministry, or Unitarian Universalist Ministry for Earth) in order to gain access to their handbooks for greening congregations or curricula to conduct classes about faith-based environmentalism.

Instrumentalism or pragmatism appears to guide the alliance building among REMOs as each party enters the coalition willing to exchange some service or resource in exchange for something they need, especially if a REMO's primary audience supports such efforts.[29] For example, VEJ and Michigan Interstate Power and Light (MIPL) work closely together on a number of projects that are mutually beneficial. Patricia Gillis runs new member orientations for the MIPL, and Father Charles Morris, its executive director, serves on VEJ's board and helps connect Michigan Catholics to VEJ. Hazon has provided funding throughout the first decade of the twenty-first century to the Shalom Center; in return, Rabbi Arthur Waskow (an esteemed leader within American Judaism) has spoken at Hazon's events, and both groups promote each other's events and programs. All of the state IPL leaders I interviewed commented on the continuing relationships they have with both the Regeneration Project (the parent REMO) and other state IPLs. Georgia IPL's Hinman notes, "We do share a lot of resources. I know, for instance, our Advent and Chanukah kits have been used by a lot of the other states. We'll direct people to other state IPL websites for resources that we don't have. We do collaborate on projects, to some extent—things like the Shop IPL and that EPA grant and everything." The EEN remains one of the members of the National Religious Partnership for the Environment not only because of shared interests and the autonomy the coalition allows, but also because the NRPE raises money for them and thus provides important financial resources they would not be able to gather otherwise.

The accounts of REMOs' search for resources follows some research on coalition formation that shows how resource-strapped groups seek out alliances in order to enhance their chances of realizing their goals.[30] Hence, it appears that REMOs with limited resources create and maintain interorganizational ties to secure the financial, human, and ideological resources that will ensure survival, and to maintain their particular brand of niche activism. Since most REMOs do not have the staff or financial wherewithal to effectively develop the educational materials they need to mobilize their audiences, they must borrow from one another. While many of these relationships are with like-minded REMOs, there are quite a few relationships that cross the lines of denomination and religious tradition. Hinman, from Georgia Interfaith Power and Light, notes that even though evangelicals do not partner with her organizations in any formal way, especially in terms of joint actions, they frequently ask for her greening and holiday kits.

REMOs' pragmatic concerns about survival may help explain why they are much more likely to enter into information-exchange relationships with a wide variety of partners. Dialogue or exchanging ideas and information is relatively safe, especially when the information focuses on environmental issues (for example, facts about climate change or suggestions for saving energy in a church). It does not require the REMO to compromise on its religious agenda, its core values and beliefs, or commit to a line of action that would make it lose credibility in the eyes of its audience (for example, a Catholic REMO joining a protest to support contraception or zero population growth). And as evinced in the last chapter, since REMO leaders are adept at borrowing ideas and fitting them into their own traditions, entering into information-exchange relationships does not pose a serious threat to their religious identity. However, joint actions require greater coordination, time, and monetary commitments than information-exchange relationships, and many REMOs voice a fear that such ties will make them appear illegitimate to their constituencies. The potential loss of "moral legitimacy" because the REMO has selected an illegitimate partner or decided on an unacceptable course of action tends to outweigh the possible rewards of joint action.[31] Thus pragmatism tends to be jettisoned when REMOs make decisions about working on common protest events or other movement campaigns.

Forging Partnerships across the Sacred-Secular Boundary

Many of the same issues that encourage or prevent alliances between REMOs—concerns about shared goals and culture, about boundary maintenance and legitimacy—also influence the formation of coalitions with their secular counterparts. REMOs work with sixty-eight secular environmental movement organizations (SEMOs) that range from national giants such as the Sierra Club to small state-based groups such as the Verde Coalition in New Mexico. Although REMOs average about two relationships with secular groups, this number hides the unequal distribution of these ties. Just over half of all REMO-SEMO ties (both information exchange and joint action) involve ten organizations, and roughly three-quarters of all connections involve just fifteen REMOs (fewer than 25 percent of all REMOs). A small number of national organizations (Sierra Club, 1Sky, Clean Water Action, and the Union of Concerned Scientists) are the most active partners.

Partnerships across the sacred-secular boundary are difficult to create and sustain because secular groups do not share the core identities and missions of religious environmentalism. REMO leaders often complain that SEMOs seem unaware of or are indifferent to REMOs' primary goal of renewing and deepening the life of faith, and therefore they do not appear to be viable partners. REMOs are not "environmental action groups," according to Rev. Katherine Jesch (UUMFE), but see their work as fundamentally about deepening faith. Sue Ellen Lowry of the Noah Alliance notes that the goal is not only to protect endangered species, but to help "recognize that we have a call to be stewards . . . and participating [in the campaign] is a way to be closer to God." The fervent commitment to their religious mission may make it difficult to partner with secular groups if they do not allow REMOs to realize such goals or if they force them to subordinate these goals to more distinctly political or secular ones.

In some cases REMOs refuse to enter into alliances with secular organizations because doing so would require the REMO to compromise on its religious commitments. This kind of mission-based conflict is evident in Rusty Pritchard's answer to my question about why the EEN publicly states that it does not work with secular environmental groups:

> A lot of the staffers [of SEMOs] will have a set of political beliefs that are pretty much at odds with some of our core attitudes toward family and issues of life and wouldn't really respect the position that Evangelicals or Orthodox Catholics would take. . . . But also given the animosity toward those kinds of groups in our community, it doesn't make much sense for us to cooperate openly. . . . As an organization, we're not at a point of being able to cooperate with them [SEMOs]. We'd like to be a little more careful with whom we appear to be allies with; we don't want to just adopt someone else's rhetoric and ideology and worldview.

The EEN's strong commitment to the values and beliefs held within Evangelicalism and their embeddedness in an audience that is unlikely to broker compromise make it nearly impossible for it to enter into partnerships with SEMOs. Pritchard's last comment also points to a second concern that REMOs have about SEMOs: co-optation. Other REMO leaders comment on the imperative to maintain clear boundaries so their religious audience is aware that the REMO is staying true to its religious mission. For example, Rev. Dennis Ormseth, executive director of Congregations Caring for Creation (C3), notes how he has struggled to not be swallowed up by the C3's secular partner, the Alliance for Sustainability: "They have a broad agenda, and we've had a hard time disentangling our efforts—not that we need to really be entirely separate, but for the purposes of public identification, we want people to understand that we're not really a front for them."

In addition, REMO leaders regularly commented on the sense that secular movement organizations only wanted a religious group to participate in a protest or join a coalition in order to either gain access to the REMO's audience or to claim they had religions' approval for their work. Katy Hinman describes this problem in a fairly even-handed manner: "In many ways secular environmental groups want to get faith groups involved because we are an untapped audience that they don't have yet and they want our numbers. And I think that's one of the reasons that folks are wary of outreach by secular environmental groups to faith communities." Jennifer Snow (PCU) is more critical:

> They are on an advocacy campaign, and their goal is to get press; their goal is to get attention; their goal is to get lots of people at their rally, and they don't really care whether or not it fits in with a given congregation's needs. I think that there is the possibility of congregations, particular clergy people because the collar can be so useful, feel that they're being used.

Snow continues by highlighting the cultural differences between religious and secular organizations and how those differences make it difficult to cooperate with one another in any meaningful or enduring manner:

> I think, as a faith-based organization, our goals are not necessarily measurable. If we want people to keep their faith and change their lives, that's not something that's measurable in the same way as getting a meeting with the mayor. And also, as someone working out of a faith-based organization, there's a sense that we may be planting seeds that we don't know how they're going to grow, and it would be counter-productive to try and control them. The idea is to let them grow as the spirit guides them. So that's not really something that makes sense to secular organizers.

Some REMO leaders report they are not sure secular environmentalists have any interest in partnering with them either. Snow, of PCU, claims that "most environmental organizations don't have any particular desire to work with communities of faith. . . . I have often come up against a real resistance and contempt for religion among secular organizations." GreenFaith has worked with a number of secular partners, and Reverend Harper, its executive director, finds that generally both sides are uncomfortable and suspicious of each other. He notes that some environmentalists think that "organized religion is part of the problem, not part of the solution." So too often there is no foundation of understanding and trust on which to build alliances.

Apart from not feeling welcome, numerous interviewees described a set of interrelated problems that limit partnerships between the religious and secular sides of the larger movement. These include different goals and ways of understanding environmental problems, incompatible operating norms, and a healthy fear of co-optation. The two sets of movement organizations seem to live in different worlds, speak different languages, and are not sure the other is a viable partner. Moreover, few activists have worked in both the religious and secular environmental fields, and thus there are few bridge figures who can help both sides understand each other and forge cooperative ties. Much of the literature that explores religious and nonreligious movements indicates that there are strong barriers to cooperation. For example, Warren found that "an organizing repertoire grounded in Christianity . . . created obstacles to extending the [faith-based] coalition to ideologically sympathetic constituencies that were either unmoved or even discomforted by this organizing schema."[32] Wood argues that religious-secular partnerships are difficult to manage. They work best when secular leaders respectfully "handle the ethical resources of religious traditions" and when the roles of religious and professional [i.e., secular] leaders are evenly balanced.[33] Rabbi Cardin of BJEN claims that the secular environmentalists she has spoken with "get a little anxious if we use words like 'caring for creation.' That's not what it's all about for them. . . . They're motivated by harsh scientific facts, like holding CO_2 to 45 parts per million, and by raising the number of bodies and the amount of money and the number of votes. They don't quite get why you need additional motivation or inspiration." Rabbi Kiener of IREJN also found that her relationship with a green energy group was difficult to manage in the early period of the partnership because the group did not want IREJN to talk about the issues in overtly religious terms:

They want me to get in there and get up on the pulpit and say "clean energy, it's here, it's real, it's working." They want me to use their buzz words. . . . And they think that I'm off message when I'm talking about the abundance of creation or what it is that really helps have a purposeful life. So there have been, and I would call those cultural differences, and that has been persistent.

Mark Jacobs of COEJL speaks more pointedly about how different organizational norms prevent closer working relationships between SEMOs and REMOs:

The environmental organizations don't know all that much about coalition building. It doesn't come naturally to the American environmental disposition, which has a kind of sense that, as an environmentalist, "I speak for all Americans . . . and so therefore, if you join me in my coalition, you're just joining the effort to speak to the public interest," even if your interest isn't the same. Environmentalists don't go and stand in the coalitions on issues related to poverty and labor and health, all sorts of things. They ask all those people to come and stand with them on those issues, but they don't understand that coalition politics really is about reciprocal relationships.

In spite of the obstacles and reservations, SEMO-REMO partnerships can work well when the boundaries between the two are maintained, and both sets of organizations are able to develop respect and trust. Dan Misleh believes the secular groups with which the Catholic Coalition on Climate Change work seek out his organization because climate change is such a huge problem that SEMOs recognize that they must recruit religious allies in order to realize their goals. But Misleh is clear that his partners must respect his religious goals and help CCCC meet them as well:

We're pretty clear with environmental groups that this has to be about people, and especially about people in poverty, not just about saving a species or something like that. Not that that's not important, but this is really where our focus is, and if you want us to help out with this, you're going to have to also toe a little bit of our line that this is about people.

Bea Moorhead from Texas IPL also notes how the boundary maintenance work she does helps make cooperation with SEMOs work well:

The reason why we're able to successfully collaborate with everybody from the ACLU to the Catholic Bishops to Planned Parenthood to the Sierra Club is that we are very assertive about drawing a line around what constitutes our constituency, what constitutes our issues. We're assertive in meetings and projects about how we work—we always work the same way—and we're clear about that with other people. . . . We're able to say "yes, we would do that" or "no, we would not do that."

Patricia Gillis (VEJ) claims that one relationship she has with a SEMO works because it respects her religious values and agenda and allows her to make faith, rather than the issue, the center point of the mobilization campaign:

I'm working with this Safer Coalition in Michigan; it's on environmental health. And one of the women said I'd like to get into some churches to do it, and that's my task is to get into churches. . . . I said, "Do you mind if we have some kind of a little devotional or prayer service before you do your thing?"

She said, "No, I don't have a problem with that." But see they would have to
be comfortable with that because . . . when I get together with faith-based
people and talk about environmental issues, I want to make the connection. I
want to have a reflection, a meditation, a prayer, something, with these folks
to help them make the connection, like the Psalms or prayers. Environmentalists would have to be okay with that; not necessarily that they would participate in it, but they would have to be okay with it.

GreenFaith's Fletcher Harper acknowledges the importance of demonstrating the particular value religious environmentalism brings to the larger
movement and specific alliances. He claims that GreenFaith is a "symboli-
cally useful partner" because it provides a new kind moral authority or legitimacy that only religion can provide, especially in the increasingly "hyper-
religious politics" that characterize American politics. In addition it is an
important "strategic partner" not only because it can "access and mobilize
certain parts of the religious community, but also because:

> we're good at what we do. . . . Just because you're a religious environmental
> group doesn't mean that you're good at what you do, and I think that there's
> a sort of quality piece that we bring to the table, as do many of them. We're a
> new part of the community. We've demonstrated that we can be a good team
> player. We helped secure and raise resources for other groups besides our own.
> We've shown that we can partner in a fair and mature way. I like to hope that
> we're a good team player, and that's certainly a part of what we aspire to do.

Several other leaders noted that religious environmentalism brings the
voice of moral authority to bear in the public sphere and a new vision of
hope that has been sorely lacking from SEMOs in recent years. Katy Hinman (Georgia IPL) offers an insightful analysis of the limits of the secular
environmental vision and the potential contribution religious environmentalism can make:

> I think there is a specific role for the faith community to play in this that's
> beyond what the secular environmental community can play. First, bring-
> ing that moral voice and really saying this is a moral issue; this isn't just an
> economic issue or an environmental issue, which you know, you can make
> the argument that environmental issues have some sort of moral underpin-
> ning. . . . But the other is that the faith community has a role to play in provid-
> ing hope and a positive vision for the future. The secular environmental com-
> munity seems constrained by the feeling that they need to be able to lay out

what steps will lead to their desired outcome. . . . And that only takes people so far. . . . Whereas the faith community is all about stepping out towards a vision that we don't know how we're going to get there; that we are taking a leap of faith that this could take a miracle, but luckily we believe in miracles. I say it kind of flippantly, but I think that's really important. I had a conversation with the former head of the Georgia Sierra Club. He said the secular environmental community is working on basically just holding back the tide, trying to keep things from getting any worse. And that is not a hopeful vision.

Successful SEMO-REMO alliances seem to emerge and endure when the secular groups come to accept the unique moral voice REMOs bring to the issues of the day. Fletcher Harper's comments about GreenFaith's contribution to the larger movement echoes Hinman's perspective that REMOs bring a vital new vision and energy to environmentalism:

> We work hard to try to cultivate a morally clear and strongly defensible position, and we speak from that, and I do think that religious leadership has a place in American culture, and I think we work hard to try to fulfill that responsibly in terms of using the public pulpit. . . . I think another piece of the picture is that we try to demonstrate to legislators and to various communities that in addition to raising red flags about problems, we try to mobilize people, at least on a small scale, to create solutions. So we're not just taking a "Chicken Little" approach to environmental problems or lamenting the fact that there are serious problems, and we're all in a heap of trouble. We're trying to say: "We can do it; here is an example of some people doing this, so let's all get with it."

Deanna Matzen from Earth Ministry claims that its secular partners not only see the value they bring to policy discussions around state environmental legislation, but they also understand that allowing the REMOs to maintain their religious identity is crucial in order to expand the base of support for the SEMOs' campaigns: "The groups [Save our Wild Salmon, The Washington Environmental Council, Partnership for a Healthy Washington] are realizing that they have to get religious people on board with the environmental agenda if change is really going to happen."

REMOs are deeply mission-driven organizations that face questions about the legitimacy of both environmentalism and working with secular groups among their constituencies and thus are constrained in their ability to enter into cooperative relations with SEMOs. Cooperation across the secular-sacred boundary is challenging and poses considerable risks. REMOs' identities are grounded in deeply held religious principles and values (that range from stewardship to the culture of life). This makes it difficult to work with secular groups who do not share the same identity or values, particularly when doing so will comprise the REMOs' religious integrity and render them illegitimate in the eyes of the religious audiences. Yet alliances between religious and secular movement organizations are facilitated when religious groups are able to pursue joint political action without having to sacrifice or compromise on their sacred core.[34] In short, embeddedness in both religious culture and religious audiences powerfully shapes the nature and scope of REMOs' cooperative endeavors.

The Limits of Cooperation

Meyer and Corrigall-Brown argue that "organizations cultivate distinct organizational identities that define them to both members and the outside world. Such identities encourage certain alliances while forestalling others. Alliances can compromise identities and can visibly associate organizations with unreliable or tainted allies."[35] This dynamic influences the structure of alliances within religious environmentalism and the practices that guide decisions about cooperation. I have tried to show how REMOs' embeddedness in more liberal religious cultures facilitates alliance formation because they value inclusivity and embrace difference, while those in more exclusive and/or conservative traditions exercise greater caution about joining alliances. This difference in orientation toward coalition formation may lie in different understandings about the importance and permeability of religious boundaries. REMOs from the liberal side seem to be less concerned about maintaining strict boundaries and more interested in opening them to other religious groups in order to accomplish their goals. Conversely, conservative groups voice much more concern about preserving or defending their religious boundaries and identity as "embattled minorities," especially when doing so is salient among their audiences.[36] REMOs' specifically religious missions and goals often pull them away from alliances because they are not necessary. Moreover, their embeddedness, and even dependence on specific religious audiences, seems to limit cooperation between REMOs and between REMOs and SEMOs. This is most evident in interviewees' comments about how particular types of partnerships—whether working with the National Council of Churches or the Sierra Club—will discredit (or at least threaten to discredit) the REMO among its constituents, erode its support, and ultimately leave the already resource-poor organization unable to effectively pursue its missions. In other words, cooperation with the wrong partner may be interpreted as a compromise with the devil that contaminates the REMO, even if it simply involves inviting a member of the Union of Concerned Scientists to speak about the science of global warming at a REMO event.

The overall field of religious environmentalism is a core-periphery network in which relations among REMOS are relatively sparse, and cooperative ties between REMOs and secular groups are even rarer. They are more likely to seek out alliances with partners who share similar theological orientations and ecological ethical positions, follow the same general set of organizational rules about organizing, or who can provide resources the REMO needs but cannot provide itself. Although information-exchange relationships are more common than joint action, both types of ties are facilitated by a shared culture and preexisting ties to each other. Acting together with groups that do not share the same religious worldview, frame, and affiliation is far less likely due to the larger risks it entails. Because collaborating on a program, joining a protest event, and engaging in some advocacy campaign are more public, they are likely more costly in terms of the potential fallout from participation among a REMO's constituents. Thus, it appears that REMOs take more care in establishing this type of relationship than they do with information-exchange ties, which are likely more private.

As Diani notes, "Any decision to set up alliances with other organizations must be carefully weighed against its potential negative impact on the actor's capacity to mobilize his own activists."[37] The fear of losing support from constituents limits joint actions between dissimilar REMOs, but also encourages activities with their own religious communities and to those national organizations that are recognized as legitimate with the tradition. Such an approach to interorganizational relationships appears to explain REMOs' isolation and leads to the core-periphery structure of the overall network.

By examining the REMO interorganizational network, we gain new insight into the ability of religious movement organizations to effect social change. The sparseness of the REMO network and the ways in which affiliation and theological commitments limit cooperative relations suggest that this nascent movement may struggle to address its targeted issues effectively, especially given the global scope of many environmental problems. Although REMOs' distinctly religious missions may make them a potentially powerful agent of change within American religion, they also severely limit their abilities to create and sustain alliances with one another and with secular movement organizations. Ultimately, REMOs' reluctance to enter into alliances and their commitment to their narrow missions may be not be effective ways to address the complexities of ecological problems, or root out the structural causes of climate change, wilderness destruction, or the health risks created by industrial agriculture. In the final chapter, I take up the future of religious environmentalism and its potential role in the larger environmental movement.

SIX

Conclusions:
Embeddedness, Strategic Choices,
and Religious Social Movements

I began this book by posing a set of questions about the new religious environmental movement: Why did it emerge nearly twenty-five years after the secular movement? How did REMOs overcome legitimation challenges and the lack of a coherent and historic environmental ethic within American religions? Why do REMOs operate in relative isolation from one another and from secular movement organizations? In short, how did this new movement emerge with few resources, in an institutional and temporal context that was not particularly opportune, and with no clear institutional or political crisis that called for a religious response? These questions defied the easy answers afforded by existing theoretical approaches to social movements. Instead, I put forth a cultural-institutional explanation that emphasizes how activists and their organizations' embeddedness in religious traditions, networks, and audiences channel, shape, and limit the emergence and development of the new movement.

The basic argument is grounded in understanding how activists' strategic choices about organizational founding, mobilization strategies, and tactical repertoires are constrained by their social ties, their commitments to particular theological and scriptural ideals, and to the sensibilities of the audiences that they hope to rally. In other words, choices about starting a REMO, creating a green religious ethic, and establishing cooperative relationships with other organizations are facilitated and constrained by the cultures and structures in which activists are located. James Jasper claims that much of the literature about social movements fails to explain how and why individuals and SMOs actually make their choices.[1] This book has been an attempt to address this problem by showing how REMO leaders' choices flow from their religious embeddedness. Although I make no claim that this new perspective is a complete theory about social movements or should replace existing approaches, I believe it is a useful conceptual tool that deepens and augments existing structural, cultural, and institutional theories. By including embeddedness in our analyses, we are more fully able to specify how cultural resources, social ties, and audience support may push activists down some pathways toward movement formation and away from others. It can help us understand how activists' work to create new movement cultures is highly patterned and constrained by particular institutional logics of appropriateness. Finally, an embeddedness perspective can show us how alliances and the overall structure of a movement field are shaped by SMOs' commitment to maintaining the integrity of their values, directed by the veto power of audiences, and organized by the meaning systems held in common by members of the movement network.[2]

Embeddedness and Movement Emergence

The long gestation period and the absence of any clearly identifiable or patterned opportunities, threats, or crises that were particularly salient to religious actors make it difficult to account for the emergence of the new movement.[3] Instead of looking for macropolitical causes to explain why the movement emerged after the 1990s, I shifted my attention to the microcultural and institutional contexts and the processes of movement formation. In chapter 2, I argued that activists' embeddedness in religious cultures provided them with the tools and the dispositions that helped them interpret ecological problems as fundamentally religious problems. More specifically, their religious identities and ways of understanding their ethical responsibilities to the world were grounded in paradigms of sin and repentance, call and response, charity and justice. Thus strip mines atop the Appalachians or oil spills in the Puget Sound came to be seen not simply as environmental disasters but as violations of the sacred mandate to care for all of creation and protect those most adversely affected by such problems. In addition, REMO founders were embedded in interpersonal and interorganizational networks whose members served as conversation partners, roles models, and sources of start-up monies.

Although some of the familiar pieces of the canonical social movement formation account (for example, catalyzing events and biographic availability) are important, I show how they are crucial in a distinctly religious way. Religious processes of conversion and awakening, rather than the rational realization of new opportunities in the political sphere, set off the founding of many REMOs. The new movement was started in part because activists believed environmental problems have underlying religious causes and, in part, because the attempt to ameliorate them was an important way to deepen one's faith and revitalize communities of faith. Ultimately, a focus on embeddedness shifts analytic attention away from the timing of movement emergence toward the particular processes of emergence. The embeddedness perspective suggests that movements may emerge as a response to the ethical and moral demands within an institutional field, rather than as a response to unsettled times and political threats, or during periods of heightened cultural and institutional creativity, as claimed by political process and cultural-institutional theories.

Embeddedness and Movement Culture

Cultural approaches to social movements examine how activists and SMOs engage in the production of meanings that will mobilize audiences by effectively framing issues and providing a new collective identity. An embeddedness perspective contributes to the study of movement culture in three ways. First, existing cultural approaches tend to emphasize how activists draw on elements (ideas, images, narratives) from their cultural repertories or borrow ideas from other institutional settings in order to craft new movement frames or meaning systems. While I have shown how REMOs are embedded in particular theological and interpretive traditions, ritual practices, and canonical texts that provide the values and principles for ethical behavior, I also argued that activists' creative endeavors are much more constrained than the literature indicates. Activists are not free to alter religious traditions, incorporate ideas from other religions, or introduce "green" historic religious rituals however they wish.[4] Creating movement cultures is guided by particular rules or "logics of appropriateness" about what legitimately may be reworked or resignified from a religious tradition and applied to the environment (for example, few Christian REMOs adopted the creation spirituality ethic because it radically diverges from the dominant interpretations of Genesis, and its biocentrism runs counter to the strong anthropocentrism of Christian ethics); how such innovation is to be accomplished such as by following the Quaker norm of consensus; and who may do so (for example, bishops and their surrogates in the case of Catholic REMOs).

Second, not all SMO cultural repertoires are equal; some religious traditions contain richer theological resources and speak more directly to the environment than others. REMOs' embeddedness in different traditions led them to craft specific strategies of innovation as they developed their green ethics. I have shown how some REMOs, such as the Jewish organizations, are embedded in religious traditions that encourage or allow for innovation or reinterpreting sacred texts, ethical precepts, or ritual practices and therefore push such REMOs to more fully use (or "mine") their traditions as they create a new green ethic. Conversely, other REMOs, such as those from conservative Protestantism, are embedded in traditions that are largely silent on and even suspicious of environmentalism and face stricter rules about how much they can reinterpret scripture or make existing ethical ideals (for example, stewardship) apply to the environment. As a result, they are more likely to rely on an innovation strategy of "bridging" that explains how environmental behavior is congruent with or an extension of the historic ethics of Protestantism. Interfaith and ecumenical REMOs are embedded in multiple traditions, and while this offers them a wider variety of cultural resources on which to draw, they must be careful about how they appropriate elements from different traditions to avoid misinterpreting a tradition or overemphasizing one particular tradition's values and ideals.[5] Both mistakes may alienate the REMOs' audience, and so they work hard to find the "common ground" or shared beliefs and values of all the traditions they seek to represent.

Third, REMOs are embedded in both formal and informal relationships of power within the larger religious field, and as Kniss and Burns argue, "the form intrareligious movements take will depend upon the polity structure within which they must operate."[6] REMOs that are tied to centralized or powerful religious authorities must follow the theological and hermeneutical stances established by religious elites and are far more constrained in their abilities to craft a green ethic. This was most evident among the Catholic and Evangelical REMOs. For the two Catholic REMOS, the key decisions about which theological and ethical values could be applied to the environment were made by bishops and others in the church hierarchy, and the REMOs largely served to convey official church teachings about the environment. Although there is not a central religious authority within American evangelicalism, REMOs that serve this constituency often found that their legitimacy depended on gaining the imprimatur from evangelical elites (megachurch clergy, publicly known parachurch leaders, evangelical scientists) and closely conforming to a pan-evangelical decision-making process based in prayer and the study of scripture. Jewish and Buddhist REMOs are embedded in decentralized and less authoritarian religious communities, while interfaith groups have decoupled themselves from the authority of national religious bodies. These three groups of REMOs face fewer sanctions from denominations and are less likely to lose legitimacy and support from their constituencies because they have not violated the official rules that govern authority relations within a given national religious organization.

Embeddedness and Movement Alliances

The literature on intramovement cooperation suggests that four factors propel SMOs to join together: shared political threats, common interests or issues, the need for resources, and preexisting interpersonal and interorganizational ties among activists. In chapter 5, I showed each of these factors shaped REMO alliances, but more importantly, I found that REMOs are wary of working with other religious and secular environmental movement organizations, despite their need for resources. The analysis demonstrates how REMOs that are strongly embedded in specific traditions and faith communities find it difficult to join alliances, when doing so may force them to compromise on their religious mission and values. This is especially the case for REMOs that believe that their supporters or potential supporters will not sanction cooperating with REMOs from other religions or with secular organizations. Thus REMOs' embeddedness in powerful audiences and incongruent systems of religious meanings limit their willingness and ability to enter into alliances with one another.

Audience pressures also influence REMOs' decisions to limit or avoid cooperating with their secular counterparts. For some REMOs, the secular movement is seen as hostile to religions or worse as a purveyor of paganism or "New Age" spirituality, which offer alternative ways of understanding nature that threaten the plausibility of Judeo-Christian teachings about creation and divine calls for stewardship. Moreover, REMOs seem wary of being too closely aligned with secular groups because they may "suffer disrepute as a result of the relations they are seen having" with them.[7] A sort of symbolic contagion or guilt by association seems to be operative among many REMO audiences, and the possibility of being discredited by working with secular movements appears to limit alliances across the sacred-secular barrier. At the same time, SEMOs and REMOs often have different values, missions, and goals; rely on different strategies; and thus have little in common on which to build relationships. Many REMO leaders voiced their concern that secular organizations were only interested in joint activities to gain access to religious markets and garner the legitimacy accorded to religion in contemporary America. Thus fear of co-optation by secular organizations is a key driver of REMOs' desire for independence, but this dynamic also flows from REMOs' and SEMOs' embeddedness in different movement cultures and audiences.

Many REMOs struggle to manage the conflicting pressures that arise if they are embedded in multiple institutional fields. For some, the demands to maintain religious boundaries restrict their ability to partner with both religious and secular groups. Others find that potential partners do not share the same stories, frames, or goals and therefore cannot find the common ground on which to build cooperative relationships. Most REMOs seem unwilling to compromise on or downplay their specific religious commitments for the sake of establishing interorganizational partnerships.[8] Even the interfaith REMOs, who are the most likely to enter into alliances, cannot always do so because their interorganizational and interpersonal ties are too homogenous; they simply do not have ties to the full range of religions in the United States (as illustrated by the case of Religious Witness for the Earth in the last chapter).[9]

The pressures to maintain religious boundaries and limit partnerships often come from REMOs' audiences. One of the unique contributions of the book is its analysis of the role audiences play in movement activities. My analyses suggest that our understanding of how movements emerge, why they adopt particular frames or repertoires of contention, and why they are effective may hinge on how well they cater to the interests and expectations of their audience members.

Embeddedness and Strategy

A focus on embeddedness also can deepen our understanding of the strategic choices SMOs make. In recent years, Jasper has argued that movement scholars need to pay more attention to how activists make strategic choices and claims that "strategic choices could be the explananda for new kinds of explanations of mobilization and conflict. We need to ask what cultural, psychological, emotional, and structural factors influence which choices are consciously made."[10] This book is a first step toward identifying how cultural and structural factors limit and direct the choices activists make about formulating goals, frames and collective identities, tactics and alliances. I have argued that the depth of activists' embeddedness in religious meaning systems, audiences, and institutional relationships shaped many of the choices they made. For example, few REMOs choose to make the preservation of wildlife a top priority because the religious traditions and audiences in which they are embedded prioritize humans above animals and the natural world. The sacred texts, theologies, and ethical systems offer few resources on which to build a credible religious case for making this the defining focus of the movement. To do so would violate the rules and assumptions within many traditions about an anthropocentric ethical focus and open up the REMOs to discrediting attacks from their constituency. In this case, "culture shapes strategy in the sense that abiding by the rules of cultural expression yields more calculable consequences than contesting them."[11]

Jasper outlines several strategic dilemmas movements face, but the questions of how and why activists make decisions to resolve these dilemmas remain underdeveloped in his accounts. The concept of embeddedness helps flesh out his strategic-choice perspective. For example, the literature suggests that REMOs should enter into cooperative relationships more often than they do. They are short of resources, have small pools of supporters, and generally have a limited ability to ameliorate the ecological problems they hope to fix, yet they work in relative isolation from other groups and are wary of joining alliances. Why? Understanding the dynamics of "the extension dilemma," in which movement leaders must weigh the potential benefits of joining coalitions or alliances against the risks of losing control over their particular SMO goals, identity, and actions, helps answer this question.[12] Many REMOs depend on an audience that does not support working with secular movement organizations or cooperating with REMOs from different faith communities. Such constraints make them unwilling to join alliances lest they lose support from their constituents. Some REMOs are embedded in religious bodies that maintain strong boundaries to outsiders, and many are entrenched in religious cultures that make it difficult to find common ground with other groups, especially secular movement organizations.[13]

Jasper and others place a great deal of attention on how interaction among movement actors or how the unfolding of action and events in the life cycle of a movement creates these strategic dilemmas and the attempts to resolve them.[14] Although I do not have the kind of fine-grained interactional data this approach requires, my interview data suggest that it is not only concrete interaction between actors in the movement field, but also the anticipated responses of REMOs' constituents to different lines of action (from joining an alliance with the Sierra Club to emphasizing justice rather than stewardship in a REMO's green ethic). REMO leaders routinely anticipate possible responses they believe church leaders, congregations, or other REMOs will make to different mobilization and protest activities, and then usually choose those that will generate the least resistance. REMOs often have "less maneuverability in the kinds of claims they can make while maintaining legitimacy" than secular movement organizations.[15] In short, strategic choices are constrained by the ways in which movement organizations are embedded in the larger religious field.

Are Religious Movements Different?

Although scholars have discussed the role religions or religious actors have played in a wide variety of social movements, they have given little attention to social movements that are religious.[16] In general, most work on this topic takes the position that religious social movements are not fundamentally different from secular movements.[17] At one level they are correct. Both nonreligious and religious movements engage in the same general set of activities—gathering resources, mobilizing support and action, framing problems, and devising solutions—and many nonreligious movements make explicit moral claims. Some scholars suggest that the contemporary environmental movement has strong spiritual or ethical dimensions, especially SEMOs' concerns with justice for those groups most adversely affected by ecological problems (that is, communities of color and the poor), and their call to rein in the overconsumption that drives Western economies.[18] Like the grassroots/antitoxins wing of the secular movement, REMOs are often founded to address some local ecological problem; they embrace a decentralized and nonprofessional organizational form, and operate with a nonpoliticized view of the world. However, REMOs do not aim to redress failures of the regulatory state, such as the antitoxins movement, nor regularly engage in direct and innovative protest activities.[19] At first glance, REMOs appear to be another type of "new social movement" (NSM). Both are decentralized and diffuse movements. Both focus on protecting particular social identities and life worlds. Both engage in noninstitutionalized politics and criticize postindustrial capitalism. Yet REMOs pursue very different goals than most new social movements. REMOs are not fighting to gain recognition and legitimacy for their own religious culture or identities as are NSMs, nor do they aim to radically democratize social relationships. Instead, as I have argued, their fundamental goal is to persuade religious audiences that they have an ethical and religious obligation to protect the environment and that doing so will help them become more authentically and fully religious.[20]

Therefore, treating both religious and secular movements as functionally equivalent misses the unique features of the former. Smith and Burns discuss this oversight in the conclusion of their review essay on religious movements and discuss why it is important to identify the particular ways in which religion may shape movement activities:

> It is difficult to find examples in the literature where religion *qua* religion is an important variable in the analysis. There is a gap in our knowledge of how variations in religious ideas and practices across different religious groups may affect their participation in broader social movements. It is reasonable to assume that how religious folk define moral problems or understand salvation from such problems will have an impact on how they define and act on social problems.[21]

Their last sentence suggests that how religious activists engage in the more general set of tasks all movements must complete may produce very different ways of organizing protest. For example, the nineteenth-century abolition and temperance movements in the United States framed the problems of slavery and alcohol use in terms of sin, and the solution in terms of confession and repentance. The movements drew on the strong emotions engendered by the confessional schema to mobilize thousands of individuals to pressure churches and the federal government to eliminate these problems.[22] Similarly, Davis and Robinson make the case that the four religiously orthodox movements they study succeeded, in part, due to their distinctly religious visions of social transformation and hope, and more importantly, to their theologies of communitarianism.[23]

I have tried to show how religious beliefs, ethical principles, sacred texts, and REMOs' overwhelming concern to deepen the divine-human relationship animate the new movement and give it a unique dynamic. The key difference between REMOs and secular movement organizations lies in the former's missions and goals. REMOs have two primary goals: first, to renew and deepen faith, and second, to address specific environmental issues. REMOs don't aim to save the whales or protest a forest's destruction purely for the sake of the animals or wilderness. The point of protecting and advocating for the environment, or in the idiom of most REMOs, "creation," is religious: to enrich one's relationship with the divine, to fulfill an individual's or community's moral obligations to God, or to more fully live an authentic life of faith. Specific issues, technical fixes, or political solutions are much less important to REMOs than to SEMOs. This overarching religious mission drives nearly every strategic choice they make. This helps us understand why REMOs focus their energies on developing new green traditions and educating religious audiences about their religious mandate to creation, rather than proposing new legislation, lobbying state and federal legislators, or organizing mass protests and citizen action campaigns (for example, letter writing and petition campaigns).

Religious environmentalists are much less focused on influencing the political sphere than their nonreligious peers. Williams contends that many religious movement organizations "are not solely focused on political change," they also may aim to meet the various religious needs and interests of their constituents.[24] Although REMOs engage in some political lobbying, it is a secondary or tertiary goal. Perhaps more importantly, the new movement can "bypass the state" and still realize its goals. The same cannot be said of many new social movements or important nonreligious movements of the late twentieth century (Black Panther, feminist, peace and conflict resolution movements) that also were not oriented toward politics. Ultimately movements, such as those listed above, that employ "identity politics" must engage the state in order to realize their goals. For example, the feminist movement had to push legislative bodies and the courts to address sexual harassment and ensure equal rights; the gay liberation movement eventually had to move beyond expanding and safeguarding lesbian and gay self-expression and enter the political sphere to end sodomy laws, employer discrimination, and win civil union and same-sex marriage rights.[25] Because its primary goal is the renewal of faith, the new movement does not have to engage in mainstream politics in the same way that other movements, even identity movements, must. And this is one of the key ways in which religious movements may be different from nonreligious ones.

REMOs also frame environmental issues differently than secular organizations. Time and again, REMO leaders contrasted their own messages of hope with the "doom-and-gloom" scenarios and discourse of secular environmentalists. Their frame did not contain the usual diagnosis of the situation in which blame is laid on actors outside of the movement (for example, extractive resource industries), nor was it anchored in the construction of enemies or the creation of an "us vs. them" identity that is so common within the nonreligious environmental movement (and many other secular movements as well). REMOs' reluctance to create an adversarial frame stems from their embeddedness in religious communities that are often ideologically and political heterogeneous. Taking a moderate, nonpartisan, inclusive approach to framing and organizing helps REMOs avoid alienating potential supporters. REMOs are more likely to include themselves, their churches and synagogues, and denominations as part of the problem because their embeddedness in religious understandings of personal and collective sin compels them to see religious communities as equally responsible for contemporary ecological problems.[26] Dan Misleh, who leads the Catholic Coalition on Climate Change, told a story about a hearing he convened in Ohio with various Catholic and environmental stakeholders, one of which was Duke Energy. The vice president from the energy company commented on the CCCC's welcoming and nonadversarial approach at the end of the hearing:

> One of the great things that he said, in our time together, was it's so nice to be in a place where I'm not vilified. He basically said this is an important role for the Catholic community, to convene people to listen to all sides and to come to solutions without giving up on our principles or our values. We need to also realize that we are part of the problem. We're all consumers of electricity. So to call energy generation companies the polluters, while it might make certain sense, it's also a little bit unfair since we all purchase what they produce.

REMOs not only offer markedly different frames about the environment, but also fill their discourse with specifically religious idioms that signal how they part company with the secular movement. For example, REMOs almost universally use the term "creation" rather than "nature," "ecology," or "environment"; they speak about stewardship and spirituality far more often than they do about conservation or preservation; and they rely on references to sacred texts more often than scientific ones to make their case that religious individuals and groups must become environmentally active. Even the movement's most notable achievements are religious in nature: distributing Earth Day resources or action kits to thousands of congregations; persuading national religious leaders to sign statements about global warming, auto emissions, or energy conservation.[27]

It seems clear that the religious environmental movement is not simply a tributary of the established, nonreligious environmental movement, but is distinct from, and runs parallel to, its secular cousin. I have shown how it is not, in the words of Paul Gorman, "the environmental movement in prayer," but a movement that aims to awaken and renew religions. It is a movement that calls individuals, congregations, and national organizations to be stewards of creation. REMOs aim to persuade their religious audiences that they can only be fully and authentically religious if they integrate environmentalism into the very fabric of their religious lives. This mission to revitalize religious belief and practice by saving the planet makes the new movement different from nonreligious ones. Thus deep religious commitments drive choices about framing, tactics, and coalition formation, as well as the nature and depth of political involvement in which to engage. It is unwise to draw conclusions about religious movements on the basis of one case study, but my analyses suggest that continued research is important to identify how and why religious movements operate in similar and different ways from nonreligious movements. In particular, comparative studies that examine religious movements, purely secular movements, and movements that have important religious participants would help us understand how religion informs organizational form, mobilization strategies, repertoires of contention, and outcomes.

The Future of Religious Environmentalism

From tentative beginnings in the mid-1990s, religious environmentalism has become firmly institutionalized within American religion. About 75 percent of the original REMOs that participated in the study are still active, and another 9 percent (or six) are in abeyance. The Regeneration Project now has chapters in forty states that serve 14,000 congregations. The National Religious Partnership for the Environment and the Eco-Justice Working Group of the NCC have distributed theological and practical information to over 150,000 congregations in the United States. Most Protestant denominations now have eco-justice or environmental stewardship programs in place and have passed proenvironmental statements.

In 2002, Gary Gardner, head of the Worldwatch Institute, wrote about the growing rapprochement between prominent US environmental organizations and religious environmental groups in his short book, *Invoking the Spirit*. He spoke optimistically about the potential of this new partnership: "The quickening of religious interest in environmental issues suggests that a powerful new political alignment may be emerging, one that would greatly strengthen the effort to build a sustainable world. A strong set of common interests inspires this tentative engagement of the spiritual and sustainable communities and would appear to make them natural allies."[28] Yet the promise of the alliance and the expansion of environmentalism into religion have not materialized. Apart from 1SKY and 350.org, none of the major national organizations currently have religious partners or programs with REMOs or national religious groups.[29] REMOs have not substantially changed public policy or legislation on the environment, and there is no evidence that they have sacralized or added a moral dimension to the discourse and culture of the broader movement. Why hasn't the movement realized its promise as Gardner envisioned? First, the lack of a shared culture and a collective identity between the religious and nonreligious groups has severely hampered cooperation. REMOs' strong commitment to mobilizing and working within religious groups and their emphasis on values rather than issues strongly limits the common ground between the two movements. In addition, REMOs' reliance on sectarian or religiously specific language and symbols suggests that they have not marshaled the kind of cultural resources to effectively shape politics and the public sphere.[30] Second, both movements are at different stages in the protest cycle, and the relative youth of REMOs means that they still must spend most of their energies educating and raising support for their work within religions. Third, REMO leaders and their audiences still view the secular movement with skepticism; REMOs are concerned about being co-opted and thus diverted from their primary religious mission; their audience members are still not sure if nonreligious movement organizations are appropriate partners. Finally, there is a shortage of leaders from both movements who can effectively bridge or connect religious and nonreligious organizations. This shortage of interpersonal and interorganizational ties has made it difficult to develop the level of trust necessary to forge coalitions across the sacred-secular line.

Despite the nascent institutionalization of environmentalism, the movement has a lot work to do in order to bring the full weight of religion to bear on the host of ecological problems it aims to address. The movement is still at the stage of raising awareness about the environment among religious audiences. Until it achieves a higher degree of "cognitive liberation" within and across religions, the movement will not be able to mobilize religious audiences to act. And, the movement is hampered by the culture and ethical commitments of its audiences.[31] That is, religions (both congregations and denominations) are multipurpose organizations that are not likely to devote resources or energies to one issue, especially one that is not purely focused on alleviating human suffering and meeting human needs.

REMOs' embeddedness within the institutional field (its traditions, its audiences, its authority systems) limits the future of the movement, but at the same time, it is responsible for its achievements. In two decades a small number of resource-poor, religious nonprofit organizations have created a new ethic and set of green religious traditions, developed the infrastructure to educate and mobilize religious individuals and congregations, and have helped hundreds of congregations and religious organizations green their facilities and worship practices. According to many of my interviewees, they have reached the "low hanging fruit" (that is, individuals and groups within a particular religious community who were already proenvironment), and now the task is to persuade the remaining 80 or 90 percent that the call to environmentalism is a sacred call that demands an immediate response. How they accomplish this formidable task will depend on how adeptly RE-MOs navigate the constraints and opportunities posed by their religious embeddedness.

APPENDIX

Interfaith

1. California Interfaith Coalition for the Environment and Children's Health
2. Congregations Caring for Creation
3. Eco-Justice Collaborative
4. Faith in Place
5. FUSE (Faith United for Sustainable Energy)
6. Georgia Interfaith Power and Light
7. GreenFaith
8. Interfaith Center for Corporate Responsibility
9. Interfaith Coalition on Energy
10. Interreligious Eco-Justice Network
11. Maine Council of Churches
12. Massachusetts Interfaith Climate Action Network
13. Michigan Interfaith Power and Light
14. National Religious Partnership for the Environment
15. Neighborhood Interfaith Movement ("Sustaining creation taskforce")
16. New York Interfaith Power and Light
17. Noah Alliance
18. Orange County Interfaith Coalition for the Environment
19. Partnership for Earth Spirituality
20. Pennsylvania Interfaith Global Climate Change Campaign
21. Regeneration Project
22. Religious Witness for Earth
23. Texas Interfaith Power and Light
24. Voices for Earth Justice
25. Wisconsin Interfaith Global Climate Change Campaign

Ecumenical

26. Alternatives for Simple Living
27. Chesapeake Covenant Congregations
28. Christians for the Mountain
29. Earthcare

30. Earth Ministry
31. Eco-Justice Ministries
32. Ecumenical Ministries of Oregon (Interfaith Network for Earth Concerns)
33. Ezekiel 34 Initiative
34. Floresta
35. Genesis Covenant
36. Grosse Ile Congregations for the Detroit River
37. Jesus People against Pollution
38. LEAF
39. National Council of Churches
40. New Community Project
41. Network Alliance of Congregations Caring for the Earth (formerly the North American Coalition for Christianity and Ecology)
42. Prairie Stewardship Network
43. Progressive Christians Uniting

Single Traditions
EVANGELICAL

44. A Rocha, USA
45. Restoring Eden
46. Evangelical Environmental Network
47. Southern Baptist Environmental and Climate Initiative

MAINLINE AND LIBERAL PROTESTANT

48. Earth Covenant Ministry
49. Presbyterians for Restoring Creation
50. Quaker Earthcare Witness
51. Unitarian Universalist Ministry for Earth
52. Web of Creation

CATHOLIC

53. National Catholic Rural Life Conference
54. Catholic Coalition on Climate Change

JEWISH

55. Baltimore Jewish Environmental Network
56. Coalition on the Environment and Jewish Life
57. Hazon
58. Sacred Foods Project (Aleph)
59. Shalom Center

BUDDHIST AND ECO-SPIRITUALITY

60. Earth and Spirit Council
61. Earth Sangha
62. Green Sangha
63. Sacred Earth Network

Nonparticipating Organizations

64. Association for Religion, Ecology, and Society
65. Buddha Gaia / Buddhist Social Action Project
66. Canfei Neshem
67. Center for Earth Spirituality and Rural Ministry

68. Churches Center for Land and People
69. Earth Justice Ministries
70. Eco-Justice Network
71. Environmental Ministries of Southern California
72. Environmental Partnership
73. Episcopal Environmental Network
74. Flourish
75. Indigenous Environmental Network
76. Interfaith Works
77. Network of Spiritual Progressives
78. Religious Campaign for Forest Prevention
79. Rhode Island Jewish Environmental Network
80. Shomrei Adamah
81. SPIN
81. Spiritual Alliance for Earth
83. Tangier Waterman's Association

NOTES

CHAPTER ONE

1. Histories of the American environmental movement often describe its evolution in terms of "waves." Dowie argues that, at the turn of the century, we were entering the fourth wave. See Mark Dowie, *Losing Ground: American Environmentalism at the Close of the Twentieth Century* (Cambridge, MA: MIT Press, 1995). See also Robert J. Brulle, *Agency, Democracy, and Nature: The U.S. Environmental Movement from a Critical Theory Perspective* (Cambridge, MA: MIT Press, 2000); Riley E. Dunlap and Angela G. Mertig, eds., *American Environmentalism: The U.S. Environmental Movement, 1970–1990* (Philadelphia: Taylor and Francis, 1992); Gregg Easterbrook, *A Moment on Earth: The Coming Age of Environmental Optimism* (New York: Penguin, 1995); and Robert Gottlieb, *Forcing the Spring: The Transformation of the American Environmental Movement* (Washington, DC: Island Press, 1993). None of these important histories mentions religion as a participant or contributor in any of the first four waves.

2. On religious movements as revitalization movements, see John A. Hannigan, "Social Movement Theory and the Sociology of Religion: Towards a New Synthesis," *Sociological Analysis* 52, no. 4 (1991): 329; and Robert Wuthnow, *Meaning and Moral Order: Explorations in Cultural Analysis* (Berkeley: University of California Press, 1987), 233–37.

3. For an overview of new social movement theory, see Jean L. Cohen, "Strategy or Identity: New Theoretical Paradigms and Contemporary Social Movements," *Social Research* 52, no. 4 (1985) (see the bibliography for page numbers for journal articles); Klaus Eder, "The 'New Social Movements': Moral Crusades, Political Pressure Groups, or Social Movements?" *Social Research* 52, no. 4 (1985); Claus Offe, "New Social Movements: Challenging the Boundaries of Institutional Politics," *Social Research* 52, no. 4 (1985); and Steven M. Buechler, "New Social Movement Theories," *Sociological Quarterly* 36, no. 3 (1995).

4. Fowler argues that Protestant eco-theory "speaks in a different language than that of the larger environmental movement," and one significant difference lies in the language and vision of hope that stands in stark contrast to the grim analyses of secular environmentalists. See Robert Booth Fowler, *The Greening of Protestant Thought* (Chapel Hill: University of North Carolina Press, 1995).

5. John M. Clements, Chenyang Xiao, and Aaron M. McCright, "An Examination of the 'Greening of Christianity' Thesis among Americans, 1993–2010," *Journal for the*

Scientific Study of Religion 53, no. 2 (2014): 373–91. This study also found few significant differences across denominations, although the authors note that Evangelical Protestants have become slightly more "green" than other Protestants since 1993.

6. Douglas Lee Eckberg and T. Jean Blocker, "Varieties of Religious Involvement and Environmental Concerns: Testing the Lynn White Thesis," *Journal for the Scientific Study of Religion* 28, no. 4 (1989) and "Christianity, Environmentalism, and the Theoretical Problems of Fundamentalism," *Journal for the Scientific Study of Religion* 35, no. 4 (1996); Conrad L. Kanagy and Hart M. Nelson, "Religion and Environmental Concern: Challenging the Dominant Assumptions," *Review of Religious Research* 37, no. 1 (1995); James L. Guth et al., "Faith and the Environment: Religious Beliefs and Attitudes on Environmental Policy," *American Journal of Political Science* 39, no. 2 (1995); Darren E. Sherkat and Christopher G. Ellison, "Structuring the Religion-Environmental Connection: Identifying Religious Influences on Environmental Concern and Activism," *Journal for the Scientific Study of Religion* 46, no. 1 (1997); and Michelle Wolkomir et al., "Denominational Subcultures of Environmentalism," *Review of Religious Research* 38, no. 4 (1997).

7. Neil Fligstein and Doug McAdam, "Toward a General Theory of Strategic Action Fields," *Sociological Theory* 29, no. 1 (2011): 1–26.

8. On the role of religions in various American social movements, see Ziad W. Munson, *The Making of Pro-Life Activists: How Social Movement Mobilization Works* (Chicago: University of Chicago Press, 2008); Michael P. Young, *Bearing Witness against Sin: The Evangelical Birth of the American Social Movement* (Chicago: University of Chicago Press, 2006); Sharon Erickson Nepstad, *Convictions of the Soul: Religion, Culture, and Agency in the Central American Solidarity Movement* (New York: Oxford University Press, 2004); Richard L. Wood, *Faith in Action: Religion, Race, and Democratic Organizing in America* (Chicago: University of Chicago Press, 2002); Mark R. Warren, *Dry Bones Rattling: Community Building to Revitalize American Democracy* (Princeton, NJ: Princeton University Press, 2001); Johnny E. Williams, "Linking Beliefs to Collective Action: Politicized Religious Beliefs and the Civil Rights Movement," *Sociological Forum* 17, no. 2 (2002); Christian Smith, *Disruptive Religion: The Force of Faith in Social-Movement Activism* (New York: Routledge, 1996) and *Resisting Reagan: The U.S. Central America Peace Movement* (Chicago: University of Chicago Press, 1996); and Aldon D. Morris, *The Origins of the Civil Rights Movement: Black Communities Organizing for Change* (New York: Free Press, 1984).

9. Nancy Tatom Ammerman, *Pillars of Faith: American Congregations and Their Partners* (Berkeley: University of California Press, 2005); Mark Chaves, *Congregations in America* (Cambridge, MA: Harvard University Press, 2004); and Robert Wuthnow, *Saving America: Faith-Based Services and the Future of Civil Society* (Princeton, NJ: Princeton University Press, 2004).

10. A handful of these groups were engaged only in organic farming, turning church properties into land trusts (such as some of the groups profiled in Taylor's study of green women's religious orders), or performing congregation-specific activities (e.g., establishing community-garden programs or enhancing energy efficiency in their religious buildings). See Sarah McFarland Taylor, *Green Sisters: A Spiritual Ecology* (Cambridge, MA: Harvard University Press, 2007).

11. An additional four were conducted in the spring of 2009.

12. Fowler, *Greening of Protestant Thought*, 14.

13. Wuthnow (*Meaning and Moral Order*, 66) defines a moral code as "a set of cultural elements that define the nature of commitment to a particular course of behavior."

14. Laurel Kearns, "Saving the Creation: Christian Environmentalism in the United States," *Sociology of Religion* 57, no. 1 (1996) and "Noah's Ark Goes to Washington: A Profile of Evangelical Environmentalism," *Social Compass* 44, no. 3 (1997). For an interfaith perspective on the early development of eco-theologies, see Mark A. Shibley and Jonathan L. Wiggins, "The Greening of Mainline American Religion: A Sociological Analysis of the Environmental Ethics of the National Religious Partnership for the Environment," *Social Compass* 44, no. 3 (1997) and James William Gibson, *A Reenchanted World: The Quest for a New Kinship with Nature* (New York: Metropolitan Books, 2009), 106–18.

15. Fowler, *Greening of Protestant Thought*, 77.

16. Kearns, "Saving the Creation," 58–60. Fowler, *Greening of Protestant Thought*, 76–90.

17. On eco-justice, see Shibley and Wiggins, "Greening of Mainline American Religion"; Fowler, *Greening of Protestant Thought*, 150–55; Kearns, "Saving the Creation"; and Gottlieb, *Forcing the Spring*, 153–61.

18. Kearns, "Noah's Ark Goes to Washington," 351. See also Kearns, "Saving the Creation" and Taylor, *Green Sisters*, 22–51.

19. For a brief list of scholarship during the 1990s, see Dieter T. Hessel and Rosemary Radford Ruether, eds., *Christianity and Ecology: Seeking the Well-Being of Earth and Humans* (Cambridge, MA: Harvard University Press, 2002), xxvii–xxx. This volume has numerous essays that describe the environmental ethics of various branches of Christianity.

20. The Pope's address can be found at http://www.vatican.va/holy_father/john_paul_ii /messages/peace/documents/hf_jp-ii_mes_19891208_xxiii-world-day-for-peace_en .html, accessed March 10, 2014.

21. On the importance of the "Open Letter," see Fowler, *Greening of Protestant Thought*; Roger S. Gottlieb, *A Greener Faith: Religious Environmentalism and Our Planet's Future* (New York: Oxford University Press, 2006); and Mary Evelyn Tucker, "Religion and Ecology," in *The Oxford Handbook of the Sociology of Religion*, ed. Peter B. Clarke (New York: Oxford University Press, 2009).

22. Some of these initiatives included the twenty-state Interfaith Global Climate Change Campaign; briefings and trainings for thousands of clergy and lay leaders about faith-based environmental work at the congregational level; and policy work on global warming, deforestation, and endangered species.

23. Gottlieb, *A Greener Faith*; Angela M. Smith and Simone Pulver, "Ethics-Based Environmentalism in Practice: Religious-Environmental Organizations in the United States," *Worldviews* 13, no. 2 (2009); William Somplatsky-Jarman, Walt E. Grazer, and Stan L. LeQuire, "Partnership for the Environment among U.S. Christians: Reports from the National Religious Partnership for the Environment," in *Christianity and Ecology: Seeking the Well-Being of Earth and Humans*, ed. Dieter T. Hessel and Rosemary Radford Ruether (Cambridge, MA: Harvard University Press, 2000); and Tucker, "Religion and Ecology."

24. According to Andrews and Edwards, "prefigurative action" refers to a movement strategy intended to cultivate and model alternative lifestyles. See Kenneth T. Andrews and Bob Edwards, "The Organizational Structures of Local Environmentalism," *Mobilization* 10, no. 2 (2005): 221.

25. I use the term "secular" to connote organizations that do not have an explicit religious identity or affiliation with an organized religious group. To be fair, many secular environmental groups have strong spiritual or moral roots in which nature is understood to be sacred. The roots of what Taylor calls "deep green religion" are in deep

ecology, Native American religions, and nineteenth- and twentieth-century American Romanticism. See Bron Taylor, "Earth and Nature-Based Spirituality (Part I): From Deep Ecology to Radical Environmentalism," *Religion* 31, no. 2 (2001) and "Earth and Nature-Based Spirituality (Part II): From Earth First! and Bioregionalism to Scientific Paganism and the New Age," *Religion* 31, no. 3 (2001).

26. On the tensions between the environmental movement and religions, see Katharine K. Wilkinson, *Between God and Green: How Evangelicals Are Cultivating a Middle Ground on Climate Change* (New York: Oxford University Press, 2012), 90–91; Gary Gardner, *Invoking the Spirit: Religion and Spirituality in the Quest for a Sustainable World*, Worldwatch Paper No. 164 (Washington, DC: Worldwatch Institute, 2002), 24–27; and Fowler, *Greening of Protestant Thought*, 19–20. Kearns recalls that organizers of the 1991 First People of Color Environmental Leadership Summit intentionally did not invite religions to participate because "we don't want Christ preached." See Laurel Kearns, "Ecology and the Environment," in *The Blackwell Companion to Religion and Social Justice*, ed. Michael D. Palmer and Stanley M. Burgess (Malden, MA: Wiley-Blackwell, 2012).

27. Robert Wuthnow, *The Restructuring of American Religion: Society and Faith since World War II* (Princeton, NJ: Princeton University Press, 1988); R. Stephen Warner, "Work in Progress toward a New Paradigm for the Sociological Study of Religion in the United States," *American Journal of Sociology* 98, no. 5 (1993); and Donald E. Miller, *Reinventing American Protestantism: Christianity in the New Millennium* (Berkeley: University of California Press, 1997), 1–10.

28. Kearns, "Saving the Creation" and "Noah's Ark Goes to Washington."

29. Smith and Pulver, "Ethics-Based Environmentalism in Practice," 148–49. See also Kearns, "Saving the Creation," and Samuel C. Potter, "The Pacific Northwest Forest Debate: Bringing Religion Back In?" *Worldviews* 3, no. 1 (1999): 3–32.

30. Dowie, *Losing Ground*, 63–82; and "The Death of Environmentalism: Global Warming in a Post-Environmental World," Michael Shellenberger and Ted Nordhaus, accessed May 22, 2015, http://www.thebreakthrough.org/images/Death_of _Environmentalism.pdf.

31. My approach draws on the cultural-institutionalist approach of Armstrong. See Elizabeth A. Armstrong, "Crisis, Collective Creativity, and the Generation of New Organizational Forms: The Transformation of Lesbian/Gay Organizations in San Francisco," *Social Structure and Organizations Revisited* 19 (2002) and *Forging Gay Identities: Organizing Sexuality in San Francisco, 1950–1994* (Chicago: University of Chicago Press, 2002); and Elizabeth A. Armstrong and Mary Bernstein, "Culture, Power, and Institutions: A Multi-Institutional Politics Approach to Social Movements," *Sociological Theory* 26, no. 1 (2008).

32. Steven Rathgeb Smith and Michael R. Sosin, "The Varieties of Faith-Related Agencies," *Public Administrative Review* 61, no. 6 (2001): 653. See also Elisabeth S. Clemens and Deborah Minkoff, "Beyond the Iron Law: Rethinking the Place of Organizations in Social Movement Research," in *The Blackwell Companion to Social Movements*, ed. David A. Snow, Sarah A. Soule, and Hanspeter Kriesi (Malden, MA: Blackwell, 2004). On the integration of neoinstitutionalism and social movement theories, see Gerald F. Davis et al., *Social Movements and Organization Theory* (New York: Cambridge University Press, 2005); Michael Lounsbury, Marc Ventresca, and Paul M. Hirsch, "Social Movements, Field Frames, and Industry Emergence: A Cultural-Political Perspective on US Recycling," *Socio-Economic Review* 1, no. 1 (2003); Hayagreeva Rao, Calvin Morrill, and Mayer N. Zald, "Power Plays: How Collective Movements and

Collective Action Create New Organizational Forms," *Research in Organizational Behavior* 22 (2000); and Kelly Moore, *Disrupting Science: Social Movements, American Scientists, and the Politics of the Military, 1945–1975* (Princeton, NJ: Princeton University Press, 2008).

33. Armstrong, "Crisis, Collective Creativity, and the Generation of New Organizational Forms," 364. Other cultural perspectives on social movements also emphasize the critical role that social change, social crises, or "moral shocks" play in movement emergence. For example, see Sharon Erickson Nepstad, *Convictions of the Soul*, and "Oppositional Consciousness among the Privileged: Remaking Religion in the Central America Solidarity Movement," *Critical Sociology* 33, no. 4 (2007); and Young, *Bearing Witness against Sin*.

34. Mark Granovetter, "Economic Action and Social Structure: The Problem of Embeddedness," *American Journal of Sociology* 91, no. 3 (1985): 481–510. Although most scholars apply embeddedness to economic activity, some have extended it to show how it is a useful concept to explain organizational behavior—Tina Dacin, Marc J. Ventresca, and Brent D. Beal, "The Embeddedness of Organizations: Dialogue & Directions," *Journal of Management* 25, no. 3 (1999)—and social movement mobilization—Bert Klandermans, Jojanneke van der Toorn, and Jacquelin van Stekelenburg, "Embeddedness and Identity: How Immigrants Turn Grievances into Action," *American Sociological Review* 73, no. 6 (2008). See also Beth Caniglia, "Informal Alliances vs. Institutional Ties: The Effects of Elite Alliances on Environmental TSMO Networks," *Mobilization* 6, no. 1 (2001).

35. Holly J. McCammon and Nella Van Dyke. "Applying Qualitative Comparative Analysis to Empirical Studies of Social Movement Coalition Formation," in *Strategic Alliances: Coalition Building and Social Movements*, ed. Nella Van Dyke and Holly J. McCammon (Minneapolis: University of Minnesota Press, 2010), 298.

36. For examples of interpersonal ties that facilitate bridge building among movement organizations and reduce differences, see Fred Rose, *Coalitions across the Class Divide: Lessons from the Labor, Peace, and Environmental Movements* (Ithaca, NY: Cornell University Press, 2000); Joe Bandy, "Paradoxes of Transnational Civil Societies under Neoliberalism: The Coalition for Justice in the Maquiladoras," *Social Problems* 51, no. 3 (2004); and Brian K. Obach, *Labor and the Environmental Movement: A Quest for Common Ground* (Cambridge, MA: MIT Press, 2004).

37. Sharon Zukin and Paul DiMaggio, "Introduction," in *Structures of Capital: The Social Organizations of the Economy*, ed. Sharon Zukin and Paul DiMaggio (New York: Cambridge University Press, 1990), 18–20; Dacin, Ventresca, and Beal, "Embeddedness of Organizations," 324–26; and Mark A. Hager, Joseph Galaskiewicz, and Jeff A. Larson, "Structural Embeddedness and the Liability of Newness among Nonprofit Organizations," *Public Management Review* 6, no. 2 (2004): 164.

38. See Danny Miller, "The Embeddedness of Corporate Strategy: Isomorphism vs. Differentiation," *Advances in Strategic Management* 13 (1996): 287–89, and Dacin, Ventresca, and Beal, "Embeddedness of Organizations," 337, for a discussion of the intensity of organizational embeddedness.

39. See Smith and Sosin, "Varieties of Faith-Related Agencies," on tight and loose coupling of faith-related agencies and national church bodies. See Wilkinson, *Between God and Green*, on religious authority and environmental activism within American evangelicalism. See D. Michael Lindsay, "Evangelicals in the Power Elite: Elite Cohesion Advancing a Movement," *American Sociological Review* 73, no. 1 (2008): 60–82; Lyman A. Kellstedt and John C. Green, "The Politics of the Willow Creek Associa-

tion," *Journal for the Scientific Study of Religion* 42, no. 4 (2003); and Wuthnow, *Restructuring of American Religion*, for more general accounts of power and authority within conservative Protestantism.

40. Dacin, Ventresca, and Beal, "Embeddedness of Organizations," 329. See also David Dequech, "Cognitive and Cultural Embeddedness: Combining Institutional Economics and Economic Sociology," *Journal of Economic Issues* 37, no. 2 (2003), for an in-depth look at cultural embeddedness.

41. Mustafa Emirbayer and Jeff Goodwin, "Network Analysis, Culture, and the Problem of Agency," *American Journal of Sociology* 99, no. 6 (1994): 1441.

42. Francesca Polletta, "Culture and Movements," *ANNALS of the American Academy of Political and Social Science* 619 (September 2008): 85–86.

43. Polletta, "Culture and Movements," 86. Jasper makes a similar point about how the meaning systems in which movements are embedded direct their strategic choices. See James M. Jasper, "A Strategic Approach to Collective Action: Looking for Agency in Social Movement Choices," *Mobilization* 9, no. 1 (2004).

44. Polletta, "Culture and Movements," 86.

45. Greta Hsu and Michael T. Hannan, "Identities, Genres, and Organizational Forms," *Organization Science* 16, no. 5 (2005): 476.

46. Ezra Zuckerman, "The Categorical Imperative: Securities Analysts and the Illegitimacy Discount," *American Journal of Sociology* 104, no. 5 (1999): 1401–3.

47. Zukin and DiMaggio ("Introduction," 20) define political embeddedness as "the manner in which economic institutions and decisions are shaped by a struggle for power that involves economic actors and nonmarket institutions, particularly the state and social classes." See also Dacin, Ventresca, and Beal, "Embeddedness of Organizations," 330–32. They emphasize the role that the state, legal systems, and policy makers play in limiting or enabling different kinds of economic activities. My use of the concept focuses on relationships of power and the religious institutional rules and settings in which power is wielded.

48. Ruth Braunstein, Brad R. Fulton, and Richard L. Wood, "The Role of Bridging Cultural Practices in Racially and Socioeconomically Diverse Civic Organizations," *American Sociological Review* 79 (2014): 708.

CHAPTER TWO

1. Doug McAdam, *Political Processes and the Development of Black Insurgency, 1930–1970* (Chicago: University of Chicago Press, 1982); Herbert Kitschelt, "Political Opportunity Structures and Political Protest: Anti-Nuclear Movements in Four Democracies," *British Journal of Political Science* 16, no. 1 (1986); Doug McAdam, John D. McCarthy, and Mayer N. Zald, eds. *Comparative Perspectives on Social Movements: Political Opportunities, Mobilizing Structures, and Cultural Framings* (Cambridge: Cambridge University Press, 1996); David S. Meyer and Debra C. Minkoff, "Conceptualizing Political Opportunity," *Social Forces*, 82, no. 4 (2004); David S. Meyer, "Protest and Political Opportunities," *Annual Review of Sociology* 30 (2004); and David A. Snow and Sarah A. Soule, *A Primer on Social Movements* (New York: W. W. Norton, 2010), 1–108. An exemplary application of POS theory is Christian Smith, *Resisting Reagan*.

2. Doug McAdam, John D. McCarthy, and Mayer N. Zald, "Introduction: Opportunities, Mobilizing Structures, and Framing Processes—Toward a Synthetic, Comparative Perspective on Social Movements," in *Comparative Perspectives on Social Movements: Political Opportunities, Mobilizing Structures, and Cultural Framings*, ed. Doug

McAdam, John D. McCarthy, and Mayer N. Zald (Cambridge: Cambridge University Press, 1996), 8.

3. Polletta, "Culture and Movements," 86–87. The most thorough critique may be found in Jeff Goodwin and James M. Jasper, "Caught in a Winding, Snarling Vine: The Structural Bias of Political Process Theory," *Sociological Forum* 14, no. 1 (1999) and James M. Jasper, "Introduction: From Political Opportunity Structures to Strategic Action" in *Contention in Context: Political Opportunities and the Emergence of Protest*, ed. Jeff Goodwin and James M. Jasper (Stanford, CA: Stanford University Press, 2012), 1–33.

4. Jack A. Goldstone, "More Social Movements or Fewer? Beyond Political Opportunity Structure to Relational Fields," *Theory and Society* 33, no. 3–4 (2004): 356.

5. James M. Jasper, "A Strategic Approach to Collective Action: Looking for Agency in Social Movement Choices," *Mobilization* 9, no. 1 (2004): 5. See also Armstrong and Bernstein, "Culture, Power, and Institutions: A Multi-Institutional Politics Approach to Social Movements"; and Elizabeth A. Armstrong, "Crisis, Collective Creativity, and the Generation of New Organizational Forms: The Transformation of Lesbian/Gay Organizations in San Francisco," *Social Structure and Organizations Revisited* 19 (2002) and *Forging Gay Identities: Organizing Sexuality in San Francisco, 1950–1994* (Chicago: University of Chicago Press, 2002), who offer a cultural-institutional approach to movements that emerge when contractions between institutions become unmanageable and/or when institutional fields are in flux.

6. Jasper, "Introduction," 23.

7. Adam Isaiah Green, "Gay and Lesbian Liberation," in *Contention in Context: Political Opportunities and the Emergence of Protest*, ed. Jeff Goodwin and James M. Jasper (Stanford, CA: Stanford University Press, 2012), 202.

8. In addition, a number of scholars have found that SMOs are more likely to be founded during periods of relative prosperity, in which individuals and groups have deeper resource pools and are more likely to share those resources, than during economic downturns. The foundings of REMOs span both types of economic periods and thus suggest that resource availability is not as salient in this case as it has been for other movements. See Snow and Soule, *A Primer on Social Movements*, 88.

9. Robert Edwards and John McCarthy, "Resources and Social Movement Mobilization," in *The Blackwell Companion to Social Movements*, ed. David A. Snow, Sarah A. Soule, and Hanspeter Kriesi (Malden, MA: Blackwell, 2004), 136–37.

10. Polletta, "Culture and Movements"; Armstrong and Bernstein, "Culture, Power, and Institutions"; Goldstone, "More Social Movements or Fewer"; Mario Diani, "Networks and Social Movements: A Research Programme," in *Social Movements and Networks: Relational Approaches to Collective Action*, ed. Mario Diani and Doug McAdam (Oxford: Oxford University Press, 2003); Jasper, "A Strategic Approach to Collective Action"; and Goodwin and James, *Contention in Context*.

11. The concept of the "institutional entrepreneur" and the set of organizational tasks they face come from research and theorizing about the emergence of new organizational forms. See Howard E. Aldrich and C. Marlene Fiol, "Fools Rush In? The Institutional Context of Industry Creation," *Academy of Management Review* 19, no. 4 (1994); Hayagreeva Rao, "Caveat Emptor: The Construction of Nonprofit Consumer Watchdog Organizations," *American Journal of Sociology* 103, no. 4 (1998); Martin Ruef, "The Emergence of Organizational Forms: A Community Ecology Approach," *American Journal of Sociology* 106, no. 3 (2000); and Hayagreeva Rao, Calvin Morrill,

and Mayer N. Zald, "Power Plays: How Collective Movements and Collective Action Create New Organizational Forms." Clemens and Armstrong integrate these ideas about entrepreneurship and organizational emergence into their studies of politics and social movements, respectively. See Elisabeth S. Clemens, *The People's Lobby: Organizational Innovation and the Rise of Interest Group Politics in the United States, 1890–1925* (Chicago: University of Chicago Press, 1997) and Armstrong, "Crisis, Collective Creativity, and the Generation of New Organizational Forms."

12. Armstrong, "Crisis, Collective Creativity, and the Generation of New Organizational Forms," 364.

13. Rao, "Caveat Emptor," makes this important argument in his study of the emergence of consumer watchdog organizations.

14. Blee and McDowell make a similar point in their essay about social movement audiences. See Kathleen Blee and Amy McDowell, "Social Movement Audiences," *Sociological Forum* 27, no. 1 (2012).

15. This account comes from the unpublished foreword to the book *A Wild Faith: Jewish Ways into Wilderness, Wilderness Ways into Judaism* by Rabbi Mike Comins (Woodstock, VT: Jewish Lights Publishing, 2007). Mr. Savage sent me the foreword after our interview in 2007.

16. Merritt recruited supporters from among top leaders of the Southern Baptist Convention, its seminaries and colleges, and key megachurches in the denomination. In doing so he followed the general rule about working in a top-down manner. Jim Ball, executive director of the Evangelical Environmental Network, Jonathan Merritt of the Southern Baptist Climate Change Initiative, and Peter Ilyn of Restoring Eden also discussed how they and other evangelicals apply this top-down mobilization technique rather than a grassroots approach. Wilkinson also discusses this tactic in her account of Evangelical efforts to mobilize around climate change. See Katharine K. Wilkinson, *Between God and Green: How Evangelicals Are Cultivating a Middle Ground on Climate Change* (New York: Oxford University Press, 2012).

17. Mario Diani, *Green Networks: A Structural Analysis of the Italian Environmental Movement* (Edinburgh: Edinburgh University Press, 1995), 35.

18. Diani, *Green Networks*, 8, and "Networks and Social Movements," 304.

19. John L. Campbell, "Where Do We Stand? Common Mechanisms in Organizations and Social Movements Research," in *Social Movements and Organization Theory*, ed. Gerald F. Davis et al. (New York: Cambridge University Press, 2005), 63.

20. In our interview, Deb Kolodny of Aleph discussed how Rabbi Zelman Schachter-Shalomi, who started the Jewish Renewal Movement, coined the phrase "eco-kashruth" to broaden the practice of keeping kosher so that observant Jews would consider additional criteria (such as pesticide use on crops) when determining if a given food was kosher.

21. From http://www.c3mn.net/index, accessed July 18, 2007. The page has been updated, and the information about the founding and mission of C3 is no longer available.

22. "History: Restoring Eden," accessed May 15, 2013, http://restoringeden.org/about/history.

23. Ibid.

24. On Dominion theology, see Fowler, *The Greening of Protestant Thought*, 81, and Wilkinson, *Between God and Green*, 15–17.

25. Dillon describes how three different Catholic groups also reworked their tradition to

push for change within the larger church. Michele Dillon, *Catholic Identity: Balancing Reason, Faith, and Power* (Cambridge: Cambridge University Press, 1999).

26. A good introduction to recent scholarship about movements and networks is Diani, "Networks and Social Movements."

27. See www.nrpe.org for a short history of the NRPE, accessed January 22, 2007.

28. Young, *Bearing Witness against Sin*, 30–34.

29. The United States Conference of Catholic Bishops issued a statement titled "Renewing the Earth: An Invitation to Reflection and Action on the Environment" in 1991 and established a department of environmental justice in 1993. In 2001, they issued a statement on climate change: "Global Climate Change: A Plea for Dialogue, Prudence, and the Common Good."

30. Quote from www.progressivechristiansuniting.org/environmental-justice, accessed November 13, 2007. This page is no longer available. The following URL carries the most up-to-date information about the Eighth Day project: www.progressive christiansuniting.org/eighthday.html, accessed August 23, 2013.

31. Clifford Geertz, *The Interpretation of Cultures* (New York: Basic Books, 1973), 93–94. See also Mary Pattillo-McCoy, "Church Culture as a Strategy of Action in the Black Community," *American Sociological Review*, 63, no. 6 (1998): 767–84; Rhys Williams, "Constructing the Public Good: Social Movements and Cultural Resources," *Social Problems* 42, no. 1 (1995): 124–44; and Christian Smith, *Disruptive Religion*.

32. Warren, *Dry Bones Rattling*; Wood, *Faith in Action*; and Ammerman, *Pillars of Faith*.

CHAPTER THREE

1. The Ethics and Religious Liberty Commission is the public policy agency of the Southern Baptist Convention. Its mission is to help Baptist congregations "understand the moral demands of the gospel, apply Christian principles to moral and social problems and questions of public policy, and to promote religious liberty in cooperation with the churches and other Southern Baptists entities." The commission commonly has weighed in on national debates about gay marriage, abortion, and other family issues, evolution and science, prayer in public school, and the constitutional issues surrounding church and state. The quote appeared at http://erlc.com/erlc/about/, accessed July 8, 2013.

2. Other Evangelical interviewees (Peter Ilyn of Restoring Eden and Rusty Pritchard of the Evangelical Environmental Network) also discussed the generational divide within evangelicalism. Farrell found a statistically significant generation gap among evangelicals on a range of social issues, with younger evangelicals reporting more liberal attitudes. See Justin Farrell, "The Young and the Restless? The Liberalization of Young Evangelicals," *Journal for the Scientific Study of Religion* 50, no. 3 (2011). See also the essays in Brian Steensland and Phillip Goff, *The New Evangelical Engagement* (New York: Oxford University Press, 2014).

3. "Statement 3: Christian Moral Convictions and Our Southern Baptist Doctrines Demand Our Environmental Stewardship." http://www.baptistcreationcare.org/node/1, accessed April 9, 2015.

4. Lamb was the president of the ERLC from 1988 to 2013.

5. From a part of the GreenFaith website that is no longer available, accessed October 22, 2007.

6. From http://www.greenfaith.org/about/mission-and-areas-of-focus, accessed July 22, 2013.

7. Ibid.

8. From a part of the GreenFaith website that is no longer available, accessed July 30, 2007.

9. Rao, Morrill, and Zald, "Power Plays: How Collective Movements and Collective Action Create New Organizational Forms," 214, and Rao, "Caveat Emptor: The Construction of Nonprofit Consumer Watchdog Organizations."

10. The concept of the "institutional entrepreneur" and the set of organizational tasks they face come from research and theorizing about the emergence of new organizational forms. See Aldrich and Fiol, "Fools Rush In? The Institutional Context of Industry Creation"; Rao, "Caveat Emptor"; Ruef, "The Emergence of Organizational Forms: A Community Ecology Approach"; and Rao, Morrill, and Zald, "Power Plays." Clemens and Armstrong integrate these ideas about entrepreneurship and organizational emergence into their studies of politics and social movements, respectively. See Clemens, *The People's Lobby*, and Armstrong, "Crisis, Collective Creativity, and the Generation of New Organizational Forms: The Transformation of Lesbian/Gay Organizations in San Francisco."

11. Aldrich and Fiol, "Fools Rush In," 648–49, call this "sociopolitical legitimacy." See also J. W. Meyer and W. R. Scott, "Centralization and the Legitimacy Problems of the Local Government," in *Organizational Environments: Ritual and Rationality*, ed. John W. Meyer and W. Richard Scott (Beverly Hills, CA: Sage, 1983); M. C. Suchman, "Managing Legitimacy: Strategic and Institutional Approaches," *Academy of Management Review* 20 (1995); and Roy Suddaby and Royston Greenwood, "Rhetorical Strategies of Legitimacy," *Administrative Science Quarterly* 50, no. 1 (2005).

12. Steve Maguire, Cynthia Hardy, and Thomas B. Lawrence, "Institutional Entrepreneurship in Emerging Fields: HIV/AIDS Treatment Advocacy in Canada," *Academy of Management Journal* 47, no. 5 (2004): 658.

13. Ruef, "Emergence of Organizational Forms," 661.

14. Organizational theorists refer to this symbolic work as "theorizing." See Paul DiMaggio, "Interest and Agency in Institutional Theory," in *Institutional Patterns and Organizations: Culture and Environment*, ed. Lynne G. Zucker (Cambridge, MA: Ballinger, 1988); David Strang and John W. Meyer, "Institutional Conditions for Diffusion," *Theory and Society*, 22, no. 4 (1993); Royston Greenwood, Roy Suddaby, and C. R. Hinings, "Theorizing Change: The Role of Professional Associations in the Transformation of Institutional Fields," *Academy of Management Journal* 45, no. 1 (2002); W. E. Douglas Creed, Maureen Scully, and John R. Austin, "Clothes Make the Person? Tailoring of Legitimating Accounts and the Social Construction of Identity," *Organization Science* 13, no. 5 (2002); and Suddaby and Greenwood, "Rhetorical Strategies of Legitimacy." However, Rao and his colleagues rely on the social movement concept of "framing" to describe the same process. See Rao, "Caveat Emptor," and Rao, Morrill, and Zald, "Power Plays."

15. Suddaby and Greenwood, "Rhetorical Strategies of Legitimacy," 38. See also Armstrong, *Forging Gay Identities* and "Crisis, Collective Creativity, and the Generation of New Organizational Forms."

16. Rao, Morrill, and Zald, "Power Plays," 249.

17. Christopher C. Morphew and Matthew Hartley, "Mission Statements: A Thematic Analysis of Rhetoric across Institutional Type," *Journal of Higher Education* 77, no. 3 (2006), and Janet A. Weiss and Sandy Dritin Piderit, "The Value of Mission Statements in Public Agencies," *Journal of Public Administration and Theory: J-Part* 9, no. 2 (1999).

18. Numerous neoinstitutionalist scholars have written about the ways in which institutional and organizational fields constrain innovation. See Rao, Morrill, and Zald, "Power Plays"; Armstrong and Bernstein, "Culture, Power, and Institutions: A Multi-Institutional Politics Approach to Social Movements"; Suddaby and Greenwood, "Rhetorical Strategies of Legitimacy"; and Roger Friedland and Robert R. Alford, "Bringing Society Back In: Symbols, Practices, and Institutional Contradictions," in *The New Institutionalism in Organizational Analysis*, ed. Walter W. Powell and Paul J. DiMaggio (Chicago: University of Chicago Press, 1991), 232–63. See Royston Greenwood and Roy Suddaby, "Institutional Entrepreneurship in Mature Fields: The Big Five Accounting Firms," *Academy of Management Review* 49, no. 1 (2006): 27–48, for a helpful summary of the network and neoinstitutionalist literatures that examine the relationship between embeddedness and organizational innovation.

19. Chaves discusses the fine line religious leaders walk to balance innovation and conformity with the past as they seek to modify worship practices. See Mark Chaves, *Congregations in America* (Cambridge, MA: Harvard University Press, 2004), 127–65.

20. The quote is from an unpublished document titled "What Is Green Sangha?" sent to me by interviewee Trathen Heckman on December 17, 2007.

21. This move to stress ethics over belief may also reflect a larger shift away from strong denominational identities and loyalties and the declining salience of maintaining theological purity or adhering to dogmas around core issues such as communion or the resurrection. On this point, see Roger S. Gottlieb, *A Greener Faith: Religious Environmentalism and Our Planet's Future* (New York: Oxford University Press, 2006), 62.

22. Such REMOs are akin to the "network organizations" described by Powell and Polody and Page. They are "engaged in reciprocal, preferential, mutually supportive actions," pool resources, and pursue joint actions rather than each member's unique interests. See Walter W. Powell, "Neither Market Nor Hierarchy: Network Forms of Organization," *Research in Organizational Behavior* 12 (1990): 303, and Joel Polody and Karen L. Page, "Network Forms of Organization," *Annual Review of Sociology* 24 (1998).

23. From "Ten-Year Anniversary Report: What's Jewish about Protecting the Environment?" on the Coalition on the Environment and Jewish Life website (http://www.coejl.org). The report is no longer available online, accessed July 30, 2013.

24. From http://www.miipl.org, accessed Aug. 22, 2007.

25. For recent research on American congregations and locally oriented faith-based organizing, see Ammerman, *Pillars of Faith*; Chaves, *Congregations in America*; and Wood, *Faith in Action*.

26. The reference to "who we are" and "how we do things here" are from Becker's discussion of cultural models of local religious communities in her study of Oak Park, Illinois. See Penny Edgell Becker, *Congregations in Conflict: Cultural Models of Local Religious Life* (Cambridge: Cambridge University Press, 1999), 1–25.

27. I'm indebted to Rich Wood for this insight.

28. Sandra R. Levitsky, "Niche Activism: Constructing a Unified Movement Identity in a Heterogeneous Organizational Field," *Mobilization* 12, no. 3 (2007): 271–72.

29. Christians for the Mountains, "Letter of Declaration." http://www.christiansforthemountains.org/site/Topics/About/Letter_of_Declaration_May2005.html, accessed June 1, 2013.

30. Catholic Rural Life: Applying the teachings of Jesus Christ for the betterment of rural America, www.ncrlc.com/magazine-webpages/01_Rural_roots_W06.html, accessed October 11, 2007. Portions of the site are no longer available; see http://catholicrurallife.org.

31. This discussion draws on Andrews and Edwards's notion of "strategic orientation" and tactics. See Andrews and Edwards, "The Organizational Structures of Local Environmentalism," 218–21.

32. For discussions of the tactics and repertoires of contention commonly used within the nonreligious environmental movement, see Andrews and Edwards, "The Organizational Structures," and Joann Carmin, "Selecting Repertoires of Action in Environmental Movement Organizations," *Organization and Environment* 15, no. 4 (2002).

33. The term "prefigurative" refers to a type of movement strategy that entails more than simply advocating for social and political change but enacting the change via lifestyle choices as a means of modeling the kind of future society the movement hopes to achieve. Breines claims prefigurative politics aim "to create and sustain within the live practice of the movement, relationships and political forms that 'prefigure' and embody the desired society." See Wini Breines, *Community and Organization in the New Left, 1962–1968: The Great Refusal* (South Hadley, MA: J. F. Bergin, 1982), 6.

34. Chaves, *Congregations in America*, 155–56.

35. For example, COEJL tried to create a national, state-chapter organization like the Sierra Club but was unable to financially sustain an organization of this size. By the mid-2000s, COEJL was down to a skeleton national staff and did very little programming.

36. Jasper, "Introduction: From Political Opportunity Structures to Strategic Action."

CHAPTER FOUR

1. From "Not Idiots," *Eco-Justice Notes* (January 30, 2004, and May 8, 2009), on the Eco-Justice website. This site is no longer available, accessed July 7, 2009.

2. On reframing, see Wade Clark Roof, *Spiritual Marketplace: Baby Boomers and the Remaking of American Religion* (Princeton, NJ: Princeton University Press, 1999), 169–71. On frame bridging, see David A. Snow et al., "Frame Alignment Processes, Micromobilization, and Movement Participation," *American Sociological Review* 51, no. 4 (1986). On Bricolage, see John L. Campbell, "Institutional Reproduction and Change," in *The Oxford Handbook of Comparative Institutional Analysis*, ed. Glenn Morgan et al. (New York: Oxford University Press, 2010).

3. Hayagreeva Rao, Philippe Monin, and Rodolphe Durand, "Border Crossing: Bricolage and the Erosion of Categorical Boundaries in French Gastronomy," *American Sociological Review* 70, no. 6 (2005); Ted Baker and E. Nelson Reed, "Creating Something from Nothing: Resource Construction through Entrepreneurial Bricolage," *Administrative Science Quarterly* 50, no. 3 (2005); Barbara Myerhoff, *Number Our Days: A Triumph of Continuity and Culture among Jewish Old People in an Urban Ghetto* (New York: Simon and Schuster, 1978); Nurit Stadler, "Is Profane Work an Obstacle to Salvation? The Case of Ultra Orthodox (Haredi) Jews in Contemporary Israel," *Sociology of Religion* 63, no. 4 (2002).

4. Roof, *Spiritual Marketplace*, 171.

5. The literature on framing is quite extensive; see David Snow, Anne E. Tan, and Peter B. Owens, "Social Movements, Framing Processes, and Cultural Revitalization and Fabrication," *Mobilization* 18, no. 3, (2013), for an introduction to the basic types of framework.

6. Snow, Tan, and Owens, "Social Movements," 227.

7. Ann Swidler, "Culture in Action: Symbols and Strategies," *American Sociological Review* 51, no. 2 (1986).

8. Robert Wuthnow, *Meaning and Moral Order: Explorations in Cultural Analysis* (Berkeley: University of California Press, 1987).

9. On organizational fields, see Elizabeth A. Armstrong, *Forging Gay Identities: Organizing Sexuality in San Francisco, 1950–1994* (Chicago: University of Chicago Press, 2002), 9.

10. Stephen Hart, *What Does the Lord Require?: How American Christians Think about Economic Justice* (New Brunswick, NJ: Rutgers University Press, 1996). See also James K. Wellman, *Evangelical vs. Liberal: The Clash of Christian Cultures in the Pacific Northwest* (New York: Oxford University Press, 2008).

11. Kniss and Campbell note that Evangelical relief and development agencies emphasize the individual rather than systemic causes and solutions, and rely on explicitly biblicist language, while mainline groups operate with a structural orientation and rely on a broader humanistic language of justice. See Fred Kniss and David Todd Campbell, "The Effect of Religious Orientation on International Relief and Development Organizations," *Journal for the Scientific Study of Religion* 36, no. 1 (1997).

12. Jeffrey Haydu, "Business Citizenship at Work: Cultural Transposition and Class Formation in Cincinnati, 1870–1910," *American Journal of Sociology* 107, no. 6 (2002): 1432; Hayagreeva Rao and Simona Giorgi, "Code Breaking: How Entrepreneurs Exploit Cultural Logic to Generate Institutional Change," *Research in Organizational Behavior* 27 (2006): 287–89.

13. Clemens, *The People's Lobby*, 49. See also Haydu, "Business Citizenship at Work," 1426–29.

14. Rao and Giorgi, "Code Breaking," 287–89, make this argument in their analysis of how the Slow Food movement changed throughout the course of two decades.

15. Penny Edgell Becker, "Making Inclusive Communities: Congregations and the 'Problem of Race,'" *Social Problems* 45, no. 4 (1998): 467. Dillon's study of American Catholic groups advocating for women's ordination, gay rights, and a woman's right to choose illustrates how this strategy operates. Each of the three pro-change groups unearthed ideas within the historic Catholic tradition (e.g., doctrinal reflexivity) to show how it supported their visions for Catholic identity and practice that did not conform to those endorsed by the church hierarchy. See Michele Dillon, *Catholic Identity: Balancing Reason, Faith, and Power* (Cambridge: Cambridge University Press, 1999).

16. Arthur Waskow, "The Kosher Pathway: Food as God-Connection in the Life of the Jewish People," accessed May 13, 2015, https://theshalomcenter.org/node/1279.

17. From http://www.hazon.org/jewish-food-movement/overview, accessed January 15, 2014.

18. Nigel Savage, "Keeping Kosher in the Time of McDonalds and Monsanto: American Jews and the Sacred Food Movement," *National Catholic Rural Life Conference Magazine* (Winter 2006) on the National Catholic Rural Life Conference website. This page is no longer available, accessed October 11, 2007.

19. See Roof, *Spiritual Marketplace*, 167–71, for his discussion of "reframing." In short, he argues that the point of this strategy is to reinterpret religious narratives, symbols, and ideals to render the tradition more relevant and engaging in a new time period. Empirical examples of reframing include Wood's study of faith-based organizing and research on the Central American solidarity movement. See Richard L. Wood, *Faith in Action: Religion, Race, and Democratic Organizing in America* (Chicago: University of Chicago Press, 2002); Sharon Erickson Nepstad, *Convictions of the Soul: Religion,*

Culture, and Agency in the Central American Solidarity Movement (New York: Oxford University Press, 2004) and "Oppositional Consciousness among the Privileged: Remaking Religion in the Central America Solidarity Movement," *Critical Sociology* 33, no. 4 (2007); and Christian Smith, *Resisting Reagan: The U.S. Central America Peace Movement* (Chicago: University of Chicago Press, 1996), 237–49.

20. From "Treasures within the Toolbox: Seven Resources within the Jewish Tradition That Could Be Applied to Contemporary Environmentalism." During our interview, Mr. Savage preferred to talk about the programs of Hazon rather than its mission, philosophy, or framing, and when I asked he referred to his written work, which he sent me immediately after the interview.

21. Gilbert S. Rosenthal, *"Tikkun ha-Olam*: The Metamorphosis of a Concept," *Journal of Religion* 85, no. 2 (2005): 214. Rosenthal provides a detailed history of the concept and argues that it has become the dominant ethical framework for Reform, Conservative, and Reconstructionist Jews in the United States.

22. Snow et al., "Frame Alignment," 467. Many REMOs do not create frames with diagnosis, prognosis, and rationale for action, but instead try to explain why environmental action is consistent with an existing religious ethic.

23. The statement can be found at http://catholicclimatecovenant.org/catholic-teachings /Bishops, accessed January 16, 2014.

24. On Catholic social teachings, see Charles E. Curran, *Catholic Moral Theology in the United States: A History* (Washington, DC: Georgetown University Press, 2008), and Roger Haight and John Langan, "Recent Catholic Social and Ethical Teachings in Light of the Social Gospel," *Journal of Religious Ethics* 18, no. 1 (1990).

25. From http://catholicclimatecovenant.org/catholic-teachings, accessed January 16, 2014.

26. Yamane describes a similar relationship between state Catholic Conferences (i.e., political lobbying agencies of the Catholic Church) and the bishops in each state. See David Yamane, *The Catholic Church in State Politics: Negotiating Prophetic Demands and Political Realities* (Lanham, MD: Rowman and Littlefield, 2005), 69–71.

27. Sarah McFarland Taylor, *Green Sisters: A Spiritual Ecology* (Cambridge, MA: Harvard University Press, 2007), 47.

28. Keith Douglass Warner, O.F.M., "The Greening of American Catholicism: Identity, Conversion, and Continuity," *Religion and American Culture* 18, no. 1 (2008): 134.

29. The "trifecta" reference comes from Katharine K. Wilkinson, *Between God and Green: How Evangelicals Are Cultivating a Middle Ground on Climate Change* (New York: Oxford University Press, 2012), 76. Her discussion of the intraevangelicalism conflict over its emergent environmental activism can be found on pp. 66–71. See also Dubin's insightful chapter on the Christian Right's opposition to art and homosexuality (Steven C. Dubin, *Arresting Images: Impolitic Art and Uncivil Actions* [New York: Routledge, 1992]); Williams and Blackburn's analysis of operation rescue (Rhys H. Williams and Jeffrey Blackburn, "Many Are Called but Few Obey: Ideological Commitment and Activism in Operation Rescue," in *Disruptive Religion: The Force of Faith in Social-Movement Activism*, ed. Christian Smith [New York: Routledge, 1996], 167–85); Smith's overview of the moral vision of evangelicals (Smith, *American Evangelicalism*, 1998); and James Guth et al., "Theological Perspectives and Environmentalism among Religious Activists," *Journal for the Scientific Study of Religion* 32, no. 4 (1993) on the low priority the environment is given among Evangelical political activists.

30. Wilkinson, *Between God and Green*, 76.

31. Justin Farrell, "The Young and the Restless? The Liberalization of Young Evangelicals," *Journal for the Scientific Study of Religion* 50, no. 3 (2011); Joel A. Nichols, "Evangelicals and Human Rights: The Continuing Ambivalence of Evangelical Christians' Support for Human Rights," *Journal of Law and Religion* 24, no. 2 (2008–9); Robert Wuthnow, *After the Baby Boomers: How Twenty- and Thirty-Somethings Are Shaping the Future of American Religion* (Princeton, NJ: Princeton University Press, 2007), 1157–80; David Domke and Kevin Coe, *The God Strategy: How Religion Became a Political Weapon in America* (New York: Oxford University Press, 2008), 151–57; Buster G. Smith and Byron Johnson, "The Liberalization of Young Evangelicals: A Research Note," *Journal for the Scientific Study of Religion* 49, no. 2 (2010); Brian Steensland and Phillip Goff, *The New Evangelical Engagement* (New York: Oxford University Press, 2014). See also Wilford's discussion of how the Saddleback Community Church, one of the leading megachurches in the United States, has been broadening its social justice concerns to include AIDS and global health, and poverty (Justin G. Wilford, *Sacred Subdivisions: The Postsuburban Transformation of American Evangelicalism* [New York: New York University Press, 2012]).

32. From a portion of the Evangelical Environmental Network website (http://www.creationcare.org) that is no longer available, accessed January 24, 2014.

33. See Nichols, "Evangelicals and Human Rights," 654–55, on evangelicalism's "anthropology of sin." See also Wilkinson, *Between God and Green*, 66–71, for a discussion of how those opposed to Evangelical environmentalism detractors used free-market and small-government arguments to discredit the emergent movement.

34. From a page on the Evangelical Environmental Network website (http://www.creationcare.org) that is no longer available, accessed June 9, 2009.

35. According to Wilkinson, *Between God and Green*, 57–64, creation care, for Evangelical REMOs, is as central to Christian life as evangelism, worship, or charity.

36. Paul A. Djupe and Gregory W. Gwiasda, "Evangelizing the Environment: Decision Process Effects in Political Persuasion," *Journal for the Scientific Study of Religion* 49, no. 1 (2010): 73–86, at 75.

37. The quotation is from David P. Gushee, "Environmental Problem Now a Megathreat," *Creation Care* 26 (Fall 2004): 8–9.

38. Wilkinson, *Between God and Green*, 45. See also her discussion of a trickle-down strategy to recruit megachurch pastors and Christian college leaders to mobilize the grass roots (pp. 55–57). D. Michel Lindsay, *Faith in the Halls of Power: How Evangelicals Joined the American Elite* (New York: Oxford University Press, 2007) also discusses evangelicals' common practice of mobilizing elite networks to generate credibility and support for Evangelical political initiatives.

39. Although the Unitarian Universalist Association has historic roots in Christianity, it has made room for many faiths and allows individual congregations and participants to commingle and combine a wide variety of religious beliefs and practices as evinced in the following passage from the denomination's web page: "Ours is a religion with deep roots in the Christian tradition, going back to the Reformation and beyond, to early Christianity. Over the last two centuries our sources have broadened to include a spectrum ranging from Eastern religions to Western scientific humanism. Unitarian Universalists (UUs) identify with and draw inspiration from Atheism and Agnosticism, Buddhism, Christianity, Humanism, Judaism, Earth-Centered Traditions, Hinduism, Islam, and more. Many UUs have grown up in these traditions—some have grown up with no religion at all. UUs may hold one or more of those traditions'

beliefs and practice its rituals. In Unitarian Universalism, you can bring your whole self: your full identity, your questioning mind, your expansive heart." From http://www.uua.org/beliefs/welcome/index.shtml, accessed January 17, 2014.

40. GreenFaith's elaboration of its "Religious Principles of Environmental Justice" identifies these four common values as well. From a page on the GreenFaith's website (http://www.greenfaith.org) that is no longer available, accessed July 30, 2007.

41. "What Makes Food Sacred? Congregational Resources for the Abrahamic Traditions," accessed May 12, 2015, https://theshalomcenter.org/node/1279.

42. Brad Fulton and Richard L. Wood, "Interfaith Community Organizing: Emerging Theological and Organizational Challenges," *International Journal of Public Theology* 6, no. 4 (2012): 413–14.

43. Kathleen Blee and Amy McDowell, "Social Movement Audiences," *Sociological Forum* 27, no. 1 (2012): 17. See Ezra Zuckerman, "The Categorical Imperative: Securities Analysts and the Illegitimacy Discount," *American Journal of Sociology* 104, no. 5 (1999) for a discussion of how audiences exercise veto power over cultural innovation.

44. Nick Crossley, "The Social World of Networks: Combining Qualitative and Quantitative Elements in Social Network Analysis," *Sociologica* 4, no. 1 (2010): 11. Krauss notes that some of the literature about religious advocacy organizations finds that they are particularly attuned to the political attitudes and theological beliefs of the laity of the church bodies they represent and will not address issues the laity does not see as legitimate. See Rachel Krauss, "Laity, Institution, Theology, or Politics? Protestant, Catholic, and Jewish Washington Offices' Agenda Setting," *Sociology of Religion* 68, no. 1 (2007).

45. John L. Campbell, "Institutional Analysis and the Role of Ideas in Political Economy," *Theory and Society* 27, no. 3 (1998): 383.

46. On conflict over changing worship and the negative effects on individual and groups, see Stephen Ellingson, *The Megachurch and the Mainline: Remaking Religious Tradition in the Twenty-First Century* (Chicago: University of Chicago Press, 2007); Penny Edgell Becker, *Congregations in Conflict: Cultural Models of Local Religious Life* (Cambridge: Cambridge University Press, 1999); Nancy Tatom Ammerman, "Religious Identities and Religious Institutions," in *Handbook of the Sociology of Religion*, ed. Michele Dillon (Cambridge: Cambridge University Press, 2003).

47. Courtney Bender and Wendy Cadge, "Constructing Buddhism(s): Interreligious Dialogue and Religious Hybridity," *Sociology of Religion* 67, no. 3 (2006).

48. On religion and deep ecology, see David Landis Barnhill and Roger S. Gottlieb, eds., *Deep Ecology and World Religions: New Essays on Sacred Grounds* (Albany: State University of New York Press, 2001).

49. Thomas Berry, "The New Story," *Teilhard Studies* 1 (Winter 1978). See also Taylor, *Green Sisters*, 7–8, 249–51.

50. "Green Menorah Covenant Coalition: Personal, Congregational, and Public-Policy Changes to Avert Global Scorching," the Shalom Center, accessed January 30, 2014, https://theshalomcenter.org/node/1276. See also "8 Days of Hanukkah, My True Love Said to Me: 'Please Heal My Earth!,'" the Shalom Center, https://theshalomcenter .org/8-days-hanukkah-my-true-love-said-meplease-heal-my-earth, for a description of suggested climate change activities for each day of the holiday.

51. "Holiday Resources," Georgia Interfaith Power and Light, accessed January 30, 2014, http://www.gipl.org/Content/Holiday_Resources.asp.

52. "Interfaith Prayers," GreenFaith, accessed January 30, 2014, http://www.greenfaith .org/files/prayers-interfaith.pdf/at_download/file. The following REMOs offer

similar DIY green ritual resources: The Neighborhood Interfaith Movement, Prairie Stewardship Network, Earth Ministry, Web of Creation, National Council of Churches, and the state IPLs.

53. Veronique Altglas, *From Yoga to Kabbalah: Religious Exoticism and the Logics of Bricolage* (New York: Oxford University Press, 2014), 6.

54. Fiona Murray, "The Oncomouse That Roared: Hybrid Exchange Strategies as a Source of Distinction at the Boundary of Overlapping Institutions," *American Journal of Sociology* 116, no. 2 (2010): 380, notes that organizations whose culture is rigid are less likely to engage in innovation than those whose culture is more flexible or open to change.

55. The relatively decentralized nature of American Judaism weakens its institutional authority, and since these REMOs are pan-denominational, they are not subject to the authority of any single national body.

CHAPTER FIVE

1. On the notion of "brokerage" in social movements, see Roger Gould and Roberto M. Fernandez, "Structures of Mediation: A Formal Approach to Brokerage in Transaction Networks," *Sociological Methodology* 19 (1989), and Crossley, "The Social World of Networks: Combining Qualitative and Quantitative Elements in Social Network Analysis."

2. For brief summaries of the risks associated with joining a movement coalition, see David S. Meyer and Catherine Corrigall-Brown, "Coalitions and Political Context: U.S. Movements against Wars in Iraq," *Mobilization* 10, no. 3 (2005): 331, and Marie Hojnacki, "Interest Groups' Decisions to Join Alliances or Work Alone," *American Journal of Political Science* 41 (1997): 66–70. See Bandy, "Paradoxes of Transnational Civil Societies under Neoliberalism: The Coalition for Justice in the Maquiladoras," on how one transnational coalition was built around respect for diversity of its members' ideologies, goals, and activities.

3. Nella Van Dyke and Holly J. McCammon, "Introduction: Social Movement Coalition Formation," in *Strategic Alliances: Coalition Building and Social Movements*, ed. Nella Van Dyke and Holly J. McCammon (Minneapolis: University of Minnesota Press, 2010), xiv.

4. Most REMOs reported few ties to nonenvironmental religious organizations (denominations, congregations, seminaries and religious colleges, religious nonprofits, including religious political lobbying agencies at the state level), and few shared ties to the same nonenvironmental religious groups, and they did not talk about those ties in any significant manner during the interviews. Thus I have not included them in the analysis. None of the REMOs reported ties to political parties at the state or federal levels.

5. The size of nodes corresponds to the number of ties, and the shape reflects religious affiliation. Each node is labeled with the REMO's acronym.

6. Diani, *Green Networks*, 123, labels SMOs that "promote campaigns which draw support from several local organizations" much like these four core REMOs, "linking pin" organizations.

7. For a list of NCC resources, see http://www.creationjustice.org/mission-and-history.html, last accessed February 4, 2014.

8. Stephen Ellingson, Vernon Woodley, and Anthony Paik, "The Structure of Religious Environmentalism: Movement Organizations, Interorganizational Networks, and Collective Action," *Journal for the Scientific Study of Religion* 51, no. 2 (2012): 278.

9. On risks of coalition building, see Margaret Levi and Gillian H. Murphy, "Coalitions of Contention: The Case of WTO Protests in Seattle," *Political Studies* 54, no. 4 (2006): 656; Will Hathaway and David S. Meyer, "Competition and Cooperation in Social Movements," *Berkeley Journal of Sociology* 38 (1994); and Nella Van Dyke, "Crossing Movement Boundaries: Factors That Facilitate Coalition Protest by American College Students, 1930–1990," *Social Problems* 50 (2003).

10. Thomas R. Rochon and David S. Meyer, eds. *Coalitions and Political Movements: The Lessons of the Nuclear Freeze* (Boulder, CO: Lynne Reiner, 1997); Holly McCammon and Karen E. Campbell, "Allies on the Road to Victory: Coalition Formation between the Suffragists and the Women's Christian Temperance Union," *Mobilization* 7 (2002); Meyer and Corrigall-Brown, "Coalitions and Political Context"; and Suzanne Staggenborg, "Coalition Work in the Pro-Choice Movement: Organizational and Environmental Opportunities and Constraints," *Social Problems* 33 (1986); William K. Carroll and R. S. Ratner, "Master Framing and Cross-Movement Networking in Contemporary Social Movements," *Sociological Quarterly* 37, no. 4 (1996), 601–25.

11. Smith and Sosin, "The Varieties of Faith-Related Agencies."

12. Evans notes that sectarian commitments often preclude cooperation among religious groups. See John H. Evans, "Cooperative Coalitions on the Religious Right and Left: Considering the Resilience of Sectarianism," *Journal for the Scientific Study of Religion* 45, no. 2 (2006).

13. R. Kharai Brown and Roland E. Brown, "The Challenge of Religious Pluralism: The Association between Interfaith Contact and Religious Pluralism," *Review of Religious Research* 53, no. 3 (2011): 329.

14. Paul Lichterman, *Elusive Togetherness: Church Groups Trying to Bridge America's Divisions* (Princeton, NJ: Princeton University Press, 2005), 15–16, 255–56. See also Richard L. Wood, *Faith in Action: Religion, Race, and Democratic Organizing in America* (Chicago: University of Chicago Press, 2002), 232–34.

15. For example, Wood, *Faith in Action*, describes the ways in which shared operational rules help faith-based coalitions work smoothly with one another. Conversely, Roth, "Organizing on One's Own," demonstrates how conflicting ethics or different understandings of what constitutes "good politics" separated the black, Chicana, and white arms of the feminist movement during the 1960s and 1970s. Lichterman shows how different ways of building community among grassroots environmental groups prevented them from working together. See Paul Lichterman, "Piecing Together Multicultural Community: Cultural Differences in Community Building among Grass-Roots Environmentalists," *Social Problems* 42, no. 4 (1995) and *The Search for Political Community: American Activists Reinventing Commitment* (New York: Cambridge University Press, 1996).

16. Suzanne Staggenborg, "Conclusion: Research on Social Movement Coalitions," in *Strategic Alliances: Coalition Building and Social Movements*, ed. Nella Van Dyke and Holly J. McCammon (Minneapolis: University of Minnesota Press, 2010), 323.

17. Lichterman, *Elusive Togetherness*, 240, reports on a similar dynamic at work with an interfaith coalition for racial justice.

18. On this buffering strategy, see Wood, *Faith in Action*, and Roger Finke, "Innovative Returns to Tradition: Using Core Beliefs as the Foundation for Innovative Accommodation," *Journal for the Scientific Study of Religion* 43 (2004). Other REMOs use this strategy with the congregations with which they work. For example, LEAF's codirector, Pat Hudson, notes how they send out information packets about mountaintop removal and a rationale for Christians to oppose it, but don't want congregations

to feel obligated to agree and act in lockstep with the REMO. Instead, she argues, "We're saying, 'take this packet of information back to your congregation, and use it anyway you want to. It is up to you.' We're not going into a Church of God and telling them, 'This is how you have to use it.' . . . 'Take it and use it in your faith tradition, in whatever way your faith calls you to do it.'" For how the bridging strategies used by interfaith coalitions can minimize or allow for difference, see Braunstein, Fulton, and Wood, "The Role of Bridging Cultural Practices in Racially and Socioeconomically Diverse Civic Organizations."

19. "An open letter to Dobson concerning climate change," Restoring Eden, accessed February 25, 2014, http://restoringeden.org/connect/CreationVoice/2007/OpenLetter.

20. The idea of "good politics" and how different understandings limit coalition building is from Roth, "Organizing on One's Own," 105–6.

21. For a discussion of evangelicals' place in the larger American moral landscape, see Wellman, Evangelical vs. Liberal. For a discussion of the tendency of evangelicals to emphasize individual causes and solutions of social problems, see John C. Green, "Evangelical Protestants and Civic Engagement: An Overview," in A Public Faith: Evangelicals and Civic Engagement, ed. Michael Cromartie (Lanham, MD: Rowman and Littlefield, 2003).

22. Numerous scholars have identified the reluctance of conservative Protestants to enter into interfaith or ecumenical partnerships on social issues. See Charles F. Hall, "The Christian Left: Who Are They and How Are They Different from the Christian Right?" Review of Religious Research 39, no. 1 (1997); Warren, Dry Bones Rattling, 243–47; Wood, Faith in Action, 146; Ammerman, Pillars of Faith, 175–87; and Evans, "Cooperative Coalitions." For an insightful analysis of how a conservative Protestant coalition broke down due to theological differences, see Deanna Rohlinger and Jill Quadagno, "Framing Faith: Cooperation and Conflict in the U.S. Conservative Christian Political Movement," Social Movement Studies 8, no. 4 (2009). The literature tends to identify conservatives as the source of noncooperation, as does my analysis. This may be due to mainline and liberal activists placing a greater value on inclusivity and cooperation (as suggested by my analysis later in the chapter), or at least a public stance in which religious exclusivity is eschewed.

23. Lichterman, Elusive Togetherness, 235–42, shows how theological differences divided religious liberals and conservatives in an antiracist coalition, even though they agreed on the religious reasons for working against racism. Evangelicals' participation and the overall effectiveness of the coalition were limited because both parties could not agree on activities (such as interfaith prayer services), partners (Wiccans), and approaches to diagnosing and addressing the problem (individual versus collective or structural).

24. On the important ways in which audiences grant or withhold legitimacy for SMOs, see Hathaway and Meyer, "Competition and Cooperation"; Blee and McDowell, "Social Movement Audiences"; and Michael T. Heaney and Fabio Rojas, "Hybrid Activism: Social Movement Mobilization in a Multi-Movement Environment," American Journal of Sociology 119, no. 4 (2014). The term "legitimacy discount" is from Zuckerman, "The Categorical Imperative: Securities Analysts and the Illegitimacy Discount."

25. Holly J. McCammon and Nella Van Dyke, "Applying Qualitative Comparative Analysis to Empirical Studies of Social Movement Coalition Formation," in Strategic Alliances: Coalition Building and Social Movements, ed. Nella Van Dyke and Holly J. McCammon (Minneapolis: University of Minnesota Press, 2010), 298. See also Mario Diani, "Networks and Participation," in The Blackwell Companion to Social Movements,

ed. David A. Snow, Sarah A. Soule, and Hanspeter Kriesi (Malden, MA: Blackwell, 2004). On the key role bridge builders play in promoting alliances, see Bandy, "Paradoxes of Transnational Civil Societies"; Campbell, "Where Do We Stand? Common Mechanisms in Organizations and Social Movements Research"; Catherine Corrigall-Brown and David S. Meyer, "The Prehistory of a Coalition: The Role of Social Ties in Win without War," in *Strategic Alliances: Coalition Building and Social Movements*, ed. Nella Van Dyke and Holly J. McCammon (Minneapolis: University of Minnesota Press, 2010); Ranjay Gulati and Martin Gargiulo, "Where Do Interorganizational Networks Come From?" *American Journal of Sociology* 104 (1999); and Levi and Murphy, "Coalitions of Contention."

26. For research that reports on the dearth of ties between evangelicals and nonevangelicals, see Fred Van Geest, "Changing Patterns of Denominational Activity in North America: The Case of Homosexuality," *Review of Religious Research* 49, no. 2 (2007); Kellstedt and Green, "The Politics of the Willow Creek Association."

27. Ellingson, Woodley, and Paik, "The Structure of Religious Environmentalism," 278.

28. On how threats mobilize SMO alliances, see McCammon and Van Dyke, "Applying Qualitative Comparative Analysis," 296–97; Staggenborg, "Coalition Work in the Pro-Choice Movement"; and McCammon and Campbell, "Allies on the Road to Victory."

29. Patrick F. Gilham and Bob Edwards, "Legitimacy, Management, Preservation of Exchange Relationships, and the Dissolution of the Mobilization for Global Justice Campaign," *Social Problems* 58, no. 3 (2011): 439.

30. McCammon and Campbell, "Allies on the Road to Victory," 238.

31. On the importance of "moral legitimacy" for social movement organizations, see Gilham and Edwards, "Legitimacy, Management, Preservation of Exchange Relationships."

32. This quote is from Clemens and Minkoff's summary of Warren's 2001 study of faith-based community organizing. See Clemens and Minkoff, "Beyond the Iron Law: Rethinking the Place of Organizations in Social Movement Research"; and Warren, *Dry Bones Rattling*.

33. Wood, *Faith in Action*, 75.

34. Wood, *Faith in Action*, 68–75, found a similar dynamic at play in the labor and faith-based coalition he studied in Oakland. See also Vernon L. Bates, "The Decline of a New Christian Right Social Movement Organization: Opportunities and Constraints," *Review of Religious Research* 42, no. 1 (2000), and Nichols, "Evangelicals and Human Rights: The Continuing Ambivalence of Evangelical Christians' Support for Human Rights."

35. Meyer and Corrigall-Brown, "Coalitions and Political Context," 331. Polletta notes that a movement associated with an inappropriate partner may discredit it among allies and supporters by "symbolic contagion" or guilt by association. See Polletta, "Culture and Movements," 88.

36. Regarding Evangelical boundary work and identity as embattled minorities, see Christian Smith, *American Evangelicalism*. See Warren, *Dry Bones Rattling*, 243–47, for an example of boundary concerns (or lack thereof) with liberal and conservative members of one of the faith-based coalitions he studied.

37. Diani, *Green Networks*, 14.

CHAPTER SIX

1. Jasper, "A Strategic Approach to Collective Action: Looking for Agency in Social Movement Choices," 2. For a similar claim, see Polletta, "Culture and Movements," 86–87. However, McCammon offers research about how movement actors "strategically

adapt" their tactics as they read events and actors in a multiorganizational field. See Holly McCammon et al., "Becoming Full Citizens: The U.S. Women's Jury Rights Campaigns, the Pace of Reform, and Strategic Adaptation," *American Journal of Sociology*, 113, no. 4 (2008).

2. For an insightful discussion of networks and meaning structures, see Jan A. Fuhse, "The Meaning Structure of Social Networks," *Sociological Theory*, 27, no. 1 (2009).

3. Jasper makes a similar claim in his discussion of a "strategic" perspective on social movement emergence. See Jasper, "Introduction: From Political Opportunity Structures to Strategic Action," 31.

4. This insight stems from Marx's opening claim in *The Eighteenth Brumaire of Louis Bonaparte*, "Men [*sic*] make their own history, but they do not make it just as they please; they do not make it under circumstances chosen by themselves, but under circumstances directly encountered, given, transmitted from the past." Although most sociologists are familiar with and agree with Marx's claim about the constraints on human agency, too often they forget to apply this insight when they describe how movement actors engage in bricolage, transposition, or framing. See Karl Marx, *The Eighteenth Brumaire of Louis Bonaparte* (New York: International Publishers, 1987 [1869]).

5. Evans makes a similar point about why some religious prochoice frames fail to resonate with specific audiences that the SMOs were trying to mobilize. See John H. Evans, "Multi-Organizational Fields and Social Movement Organization Frame Content: The Religious Pro-Choice Movement," *Sociological Inquiry* 67, no. 4 (1997).

6. Fred Kniss and Gene Burns, "Religious Movements," in *The Blackwell Companion to Social Movements*, ed. David A. Snow, Sarah Soule, and Hanspeter Kriesi (Malden, MA: Blackwell, 2004), 705. For a discussion of how faith-based organizations may be tightly or loosely coupled to official religious authority systems, and how such coupling impacts their ability to garner resources, see Smith and Sosin, "The Varieties of Faith-Related Agencies."

7. Polletta, "Culture and Movements," 88.

8. This runs counter to Braunstein's finding that some liberal religious advocacy groups do not use explicitly religious language in order to maintain strong relations with secular political actors. See Ruth Braunstein, "Storytelling in Liberal Religious Advocacy," *Journal for the Scientific Study of Religion* 51, no. 1 (2012): 110–27.

9. An anonymous reviewer of the manuscript pointed out the dilemmas that multiple embeddedness pose for REMOs.

10. Jasper, "A Strategic Approach to Collective Action," 11. See also Jasper, "Introduction."

11. Polletta, "Culture and Movements," 91.

12. Jasper, "A Strategic Approach to Collective Action," 7. See also James M. Jasper, "The Innovation Dilemma: Some Risks of Creativity in Strategic Action," in *The Dark Side of Creativity*, ed. David H. Cropley et al. (New York: Cambridge University Press, 2010).

13. Mirola shows how differences between congregational and secular movement cultures impeded coalition formation between the movement and religious organizations. See William A. Mirola, "Religious Protest and Economic Conflict: Possibilities and Constraints on Religious Resources Mobilization and Coalitions in Detroit's Newspaper Strike," *Sociology of Religions* 64, no. 4 (2003).

14. Jasper, "A Strategic Approach to Collective Action" and "Introduction"; McCammon et al., "Becoming Full Citizens"; and Marshall Ganz, *Why David Sometimes Wins: Leadership, Organization, and Strategy in the California Farm Worker Movement* (New York: Oxford University Press, 2009).

15. Jaime Kucinskas, "The Unobtrusive Tactics of Religious Movements," *Sociology of Religion* 75, no. 4 (2014): 539–38.

16. Some notable exceptions include Christian Smith, *Disruptive Religion*; Nepstad, *Convictions of the Soul*; Young, *Bearing Witness against Sin*; and Nancy J. Davis and Robert V. Robinson, *Claiming Society for God: Religious Movements and Social Welfare* (Bloomington: Indiana University Press, 2012).

17. See Christian Smith, *Disruptive Religion*, especially his opening chapter, for examples of "religion as resource provider." See also Morris, *The Origins of the Civil Rights Movement*; and Doug McAdam, *Political Process and the Development of Black Insurgency, 1930–1970* (Chicago: University of Chicago Press, 1982). For works that treat religious and secular movements as fundamentally the same, see Meyer Zald, "Theological Crucibles: Social Movements in and of Religion," *Review of Religious Research* 23, no. 1 (1982); John A. Hannigan, "Social Movement Theory and the Sociology of Religion: Towards a New Synthesis," *Sociological Analysis* 52, no. 4 (1991); James Beckford, "Social Movements as Religious Phenomena," in *The Blackwell Companion to the Sociology of Religion*, ed. Richard K. Fenn (Malden, MA: Blackwell, 2001); and Davis and Robinson, *Claiming Society for God*, 149–50.

18. See Gardner, *Invoking the Spirit*, 5; Dowie, *Losing Ground*, 205–58; and John S. Dryzek, *The Politics of the Earth: Environmental Discourses* (New York: Oxford University Press, 1997), 155–71.

19. For descriptions of the grassroots and antitoxins movement, see Phil Brown, "The Toxic Waste Movement: A New Type of Activism," *Society and Natural Resources* 7, no. 3 (1994): 269–87; Sherry Cable and Michael Benson, "Acting Locally: Environmental Injustice and the Emergence of Grass-Roots Environmental Organizations," *Social Problems* 40, no. 4 (1993): 464–77; and Nicholas Freudenberg and Carol Steinsapir, "Not in Our Backyards: The Grassroots Environmental Movement," in *American Environmentalism: The U.S. Environmental Movement, 1970–1990*, ed. Riley E. Dunlap and Angela G. Mertig (Philadelphia: Taylor and Francis, 1992), 27–37.

20. On new social movements, see Eder, "The 'New Social Movements': Moral Crusades, Political Pressure Groups, or Social Movements?"; Offe, "New Social Movements: Challenging the Boundaries of Institutional Politics"; and Hank Johnston, Enrique Larana, and Joseph R. Gusfield, "Identities, Grievances, and New Social Movements," in *New Social Movements: From Ideology to Identity*, ed. Enrique Larana, Hank Johnston, and Joseph R. Gusfield (Philadelphia: Temple University Press, 1994), 3–35.

21. Kniss and Burns, "Religious Movements," 709.

22. Young, *Bearing Witness against Sin*.

23. David and Robinson, *Claiming Society for God*, 10–31.

24. Rhys Williams, "Religious Social Movements in the Public Sphere: Organization, Ideology, and Activism," in *Handbook of the Sociology of Religion*, ed. Michele Dillon (Cambridge: Cambridge University Press, 2003), 328.

25. Davis and Robinson, *Claiming Society of God*, 149–50, argue that bypassing the state is not unique to religious movements and briefly identify several nonreligious movements that use this tactic. I have suggested that the movements they discuss ultimately had to turn to politics in order to accomplish their goals. The term "identity politics" is from Armstrong, "Crisis, Collective Creativity, and the Generation of New Organizational Forms: The Transformation of Lesbian/Gay Organizations in San Francisco," 371–76.

26. Kucinskas, "The Unobtrusive Tactics of Religious Movements," also notes that reli-

gious movements may not see the world in zero-sum, competitive, or confrontational terms and instead aim to realize their goals through consensus-building tactics.

27. The following websites, which no longer exist, offered summaries of these achievements: http://www.creationjustice.org/about/history.php (accessed February 4, 2014) and http:/www.nrpe.org/whatistheparntership/timeline.htm (accessed February 11, 2008).

28. Gardner, *Invoking the Spirit*, 7–8.

29. I searched websites of the Sierra Club, National Wildlife Federation, Wilderness Society, Environmental Defense Fund, Izaak Walton League, Audubon Society, Friends of the Earth, National Resources Defense Council, Environmental Policy Institute, and Greenpeace for signs of partnerships or joint action with religious groups and found very little evidence that these groups worked with REMOs. News stories on their websites that mentioned religion tended to be from the late 1990s and early 2000s. The Sierra Club had an official program and dedicated staff member to cultivate religious partners, but that program was closed by the end of the decade. See http:/www.1sky.org/learn/allies for a list that includes a significant number of REMOs.

30. The notion of using widely shared cultural idioms as a way of exercising cultural power in the public sphere is from the work of N. J. Demerath III and Rhys H. Williams, *A Bridging of Faiths: Religion and Politics in a New England City* (Princeton, NJ: Princeton University Press, 1992), 170.

31. The term "cognitive liberation" and its importance in mobilizing protest are discussed by McAdam, *Political Process*, as a key step in his "political process model."

BIBLIOGRAPHY

Aldrich, Howard E., and C. Marlene Fiol. "Fools Rush In? The Institutional Context of Industry Creation." *Academy of Management Review* 19, no. 4 (1994): 645–70.

Altglas, Veronique. *From Yoga to Kabbalah: Religious Exoticism and the Logics of Bricolage*. New York: Oxford University Press, 2014.

Ammerman, Nancy Tatom. *Pillars of Faith: American Congregations and Their Partners*. Berkeley: University of California Press, 2005.

———. "Religious Identities and Religious Institutions." In *Handbook of the Sociology of Religion*, edited by Michele Dillon, 207–24. Cambridge: Cambridge University Press, 2003.

Andrews, Kenneth T., and Bob Edwards. "The Organizational Structures of Local Environmentalism." *Mobilization* 10, no. 2 (2005): 213–34.

Armstrong, Elizabeth A. "Crisis, Collective Creativity, and the Generation of New Organizational Forms: The Transformation of Lesbian/Gay Organizations in San Francisco." *Social Structure and Organizations Revisited* 19 (2002): 361–95.

———. *Forging Gay Identities: Organizing Sexuality in San Francisco, 1950–1994*. Chicago: University of Chicago Press, 2002.

Armstrong, Elizabeth A., and Mary Bernstein. "Culture, Power, and Institutions: A Multi-Institutional Politics Approach to Social Movements." *Sociological Theory* 26, no. 1 (2008): 74–99.

Bainbridge, William Sims. "Science and Religion." In *The Oxford Handbook of the Sociology of Religion*, edited by Peter B. Clarke, 303–18. New York: Oxford University Press, 2009.

Baker, Ted, and E. Nelson Reed. "Creating Something from Nothing: Resource Construction through Entrepreneurial Bricolage." *Administrative Science Quarterly* 50, no. 3 (2005): 329–66.

Bandy, Joe. "Paradoxes of Transnational Civil Societies under Neoliberalism: The Coalition for Justice in the Maquiladoras." *Social Problems* 51, no. 3 (2004): 410–31.

Barnhill, David Landis, and Roger S. Gottlieb, eds. *Deep Ecology and World Religions: New Essays on Sacred Grounds*. Albany: State University of New York Press, 2001.

Bates, Vernon L. "The Decline of a New Christian Right Social Movement Organization: Opportunities and Constraints." *Review of Religious Research* 42, no. 1 (2000): 19–40.

Becker, Penny Edgell. *Congregations in Conflict: Cultural Models of Local Religious Life*. Cambridge: Cambridge University Press, 1999.

———. "Making Inclusive Communities: Congregations and the 'Problem of Race.'" *Social Problems* 45, no. 4 (1998): 451–72.

Beckford, James. "Social Movements as Religious Phenomena." In *The Blackwell Companion to the Sociology of Religion*, edited by Richard K. Fenn, 229–48. Malden, MA: Blackwell, 2001.

Bender, Courtney, and Wendy Cadge. "Constructing Buddhism(s): Interreligious Dialogue and Religious Hybridity." *Sociology of Religion* 67, no. 3 (2006): 229–47.

Berry, Thomas. "The New Story." *Teilhard Studies* 1 (Winter 1978).

Blee, Kathleen, and Amy McDowell. "Social Movement Audiences." *Sociological Forum* 27, no. 1 (2012): 1–20.

Braunstein, Ruth. "Storytelling in Liberal Religious Advocacy." *Journal for the Scientific Study of Religion* 51, no. 1 (2012): 110–27.

Braunstein, Ruth, Brad R. Fulton, and Richard L. Wood. "The Role of Bridging Cultural Practices in Racially and Socioeconomically Diverse Civic Organizations." *American Sociological Review* 79 (2014).

Breines, Wini. *Community and Organization in the New Left, 1962–1968: The Great Refusal.* South Hadley, MA: J. F. Bergin, 1982.

Brown, Phil. "The Toxic Waste Movement: A New Type of Activism." *Society and Natural Resources* 7, no. 3 (1994): 269–87.

Brown, R. Kharai, and Roland E. Brown. "The Challenge of Religious Pluralism: The Association between Interfaith Contact and Religious Pluralism." *Review of Religious Research* 53, no. 3 (2011): 329.

Brulle, Robert J. *Agency, Democracy, and Nature: The U.S. Environmental Movement from a Critical Theory Perspective.* Cambridge, MA: MIT Press, 2000.

Buechler, Steven M. "New Social Movement Theories." *Sociological Quarterly* 36, no. 3 (1995): 441–64.

Cable, Sherry, and Michael Benson, "Acting Locally: Environmental Injustice and the Emergence of Grass-Roots Environmental Organizations." *Social Problems* 40, no. 4 (1993): 464–77.

Campbell, John L. "Institutional Analysis and the Role of Ideas in Political Economy." *Theory and Society* 27, no. 3 (1998): 377–409.

———. "Institutional Reproduction and Change." In *The Oxford Handbook of Comparative Institutional Analysis*, edited by Glenn Morgan, John L. Campbell, Colin Crouch, Ove K. Pedersen, and Richard Whitley, 87–115. New York: Oxford University Press, 2010.

———. "Mechanisms of Evolutionary Change in Economic Governance: Interaction, Interpretation, and Bricolage." In *Evolutionary Economics and Path Dependence*, edited by Lars Magnusson and Jan Ottosson, 10–31. Cheltenham, UK: Edward Elgar, 1997.

———. "Where Do We Stand? Common Mechanisms in Organizations and Social Movements Research." In *Social Movements and Organization Theory*, edited by Gerald F. Davis, Doug McAdam, W. Richard Scott, and Mayer N. Zald, 41–68. New York: Cambridge University Press, 2005.

Caniglia, Beth. "Informal Alliances vs. Institutional Ties: The Effects of Elite Alliances on Environmental TSMO Networks." *Mobilization* 6, no. 1 (2001): 37–54.

Carmin, Joann. "Selecting Repertoires of Action in Environmental Movement Organizations." *Organization and Environment* 15, no. 4 (2002): 365–88.

Carroll, William K., and R. S. Ratner. "Master Framing and Cross-Movement Networking in Contemporary Social Movements." *Sociological Quarterly* 37, no. 4 (1996): 601–25.

Catholic Climate Covenant: Care for Creation; Care for the Poor. Accessed January 16, 2014. http://catholicclimatecovenant.org.

Catholic Rural Life: Applying the Teachings of Jesus Christ for the Betterment of Rural America. Accessed October 11, 2007. http://catholicrurallife.org.

Chaves, Mark. *Congregations in America*. Cambridge, MA: Harvard University Press, 2004.

Christians for the Mountains. "Letter of Declaration." Accessed June 1, 2013. http://www.christiansforthemountains.org/site/Topics/About/Letter_of_Declaration_May2005.html.

Clemens, Elisabeth S. *The People's Lobby: Organizational Innovation and the Rise of Interest Group Politics in the United States, 1890–1925*. Chicago: University of Chicago Press, 1997.

Clemens, Elisabeth S., and Deborah Minkoff. "Beyond the Iron Law: Rethinking the Place of Organizations in Social Movement Research." In *The Blackwell Companion to Social Movements*, edited by David A. Snow, Sarah A. Soule, and Hanspeter Kriesi, 155–69. Malden, MA: Blackwell, 2004.

Clements, John M., Chenyang Xiao, and Aaron M. McCright. "An Examination of the 'Greening of Christianity' Thesis among Americans, 1993–2010." *Journal for the Scientific Study of Religion* 53, no. 2 (2014): 373–91.

Cohen, Jean L. "Strategy or Identity: New Theoretical Paradigms and Contemporary Social Movements." *Social Research* 52, no. 4 (1985): 663–716.

Corrigall-Brown, Catherine, and David S. Meyer. "The Prehistory of a Coalition: The Role of Social Ties in Win without War." In *Strategic Alliances: Coalition Building and Social Movements*, edited by Nella Van Dyke and Holly J. McCammon, 3–22. Minneapolis: University of Minnesota Press, 2010.

Creation Justice Ministries. Accessed February 4, 2014. http://www.creationjustice.org/mission-and-history.html.

Creed, W. E. Douglas, Maureen Scully, and John R. Austin. "Clothes Make the Person? Tailoring of Legitimating Accounts and the Social Construction of Identity." *Organization Science* 13, no. 5 (2002): 475–96.

Cromartie, Michael, ed. *A Public Faith: Evangelicals and Civic Engagement*. Lanham, MD: Rowman and Littlefield, 2003.

Cropley, David H., Arthur J. Cropley, James C. Kaufman, and Mark A. Runco. *The Dark Side of Creativity*. New York: Cambridge University Press, 2010.

Crossley, Nick. "The Social World of Networks: Combining Qualitative and Quantitative Elements in Social Network Analysis." *Sociologica* 4, no. 1 (2010): 1–33.

Curran, Charles E. *Catholic Moral Theology in the United States: A History*. Washington, DC: Georgetown University Press, 2008.

Dacin, M. Tina, Marc J. Ventresca, and Brent D. Beal. "The Embeddedness of Organizations: Dialogue & Directions." *Journal of Management* 25, no. 3 (1999): 317–56.

Davis, Gerald F., Doug McAdam, W. Richard Scott, and Mayer N. Zald. *Social Movements and Organization Theory*. New York: Cambridge University Press, 2005.

Davis, Nancy J., and Robert V. Robinson. *Claiming Society for God: Religious Movements and Social Welfare*. Bloomington: Indiana University Press, 2012.

Demerath, N. J., III, and Rhys H. Williams. *A Bridging of Faiths: Religion and Politics in a New England City*. Princeton, NJ: Princeton University Press, 1992.

Dequech, David. "Cognitive and Cultural Embeddedness: Combining Institutional Economics and Economic Sociology." *Journal of Economic Issues* 37, no. 2 (2003): 461–70.

Diani, Mario. *Green Networks: A Structural Analysis of the Italian Environmental Movement*. Edinburgh: Edinburgh University Press, 1995.

———. "Networks and Participation." In *The Blackwell Companion to Social Movements*, edited by David A. Snow, Sarah A. Soule, and Hanspeter Kriesi, 339–59. Malden, MA: Blackwell, 2004.

———. "Networks and Social Movements: A Research Programme." In *Social Movements and Networks: Relational Approaches to Collective Action,* edited by Mario Diani and Doug McAdam, 299–319. Oxford: Oxford University Press, 2003.

Diani, Mario, and Doug McAdam. *Social Movements and Networks: Relational Approaches to Collective Action.* Oxford: Oxford University Press, 2003.

Dillon, Michele. *Catholic Identity: Balancing Reason, Faith, and Power.* Cambridge: Cambridge University Press, 1999.

DiMaggio, Paul. "Interest and Agency in Institutional Theory." In *Institutional Patterns and Organizations: Culture and Environment,* edited by Lynne G. Zucker, 3–21. Cambridge, MA: Ballinger, 1988.

Djupe, Paul A., and Gregory W. Gwiasda. "Evangelizing the Environment: Decision Process Effects in Political Persuasion." *Journal for the Scientific Study of Religion* 49, no. 1 (2010): 73–86.

Domke, David, and Kevin Coe. *The God Strategy: How Religion Became a Political Weapon in America.* New York: Oxford University Press, 2008.

Dowie, Mark. *Losing Ground: American Environmentalism at the Close of the Twentieth Century.* Cambridge, MA: MIT Press, 1995.

Dryzek, John S. *The Politics of the Earth: Environmental Discourses.* New York: Oxford University Press, 1997.

Dubin, Steven C. *Arresting Images: Impolitic Art and Uncivil Actions.* New York: Routledge, 1992.

Dunlap, Riley E., and Angela G. Mertig, eds. *American Environmentalism: The U.S. Environmental Movement, 1970–1990.* Philadelphia: Taylor and Francis, 1992.

Easterbrook, Gregg. *A Moment on Earth: The Coming Age of Environmental Optimism.* New York: Penguin, 1995.

Eckberg, Douglas Lee, and T. Jean Blocker. "Christianity, Environmentalism, and the Theoretical Problems of Fundamentalism." *Journal for the Scientific Study of Religion* 35, no. 4 (1996): 343–55.

———. "Varieties of Religious Involvement and Environmental Concerns: Testing the Lynn White Thesis." *Journal for the Scientific Study of Religion* 28, no. 4 (1989): 509–17.

Eder, Klaus. "The 'New Social Movements': Moral Crusades, Political Pressure Groups, or Social Movements?" *Social Research* 52, no. 4 (1985): 869–90.

Edwards, Robert, and John McCarthy. "Resources and Social Movement Mobilization." In *The Blackwell Companion to Social Movements,* edited by David A. Snow, Sarah A. Soule, and Hanspeter Kriesi, 116–52. Malden, MA: Blackwell, 2004.

Eighth Day Project. "What We Seek to Do." Accessed August 23, 2013. http://www.progressivechristiansuniting.org/eighthday.html.

Ellingson, Stephen. *The Megachurch and the Mainline: Remaking Religious Tradition in the Twenty-First Century.* Chicago: University of Chicago Press, 2007.

Ellingson, Stephen, Vernon Woodley, and Anthony Paik. "The Structure of Religious Environmentalism: Movement Organizations, Interorganizational Networks, and Collective Action." *Journal for the Scientific Study of Religion* 51, no. 2 (2012): 266–85.

Emirbayer, Mustafa, and Jeff Goodwin. "Network Analysis, Culture, and the Problem of Agency." *American Journal of Sociology* 99, no. 6 (1994): 1044–93.

Ethics and Religious Liberty Commission of the Southern Baptist Convention. "About the ERLC." Accessed July 8, 2013. http://erlc.com/erlc/about.

Evangelical Environmental Network. Accessed January 24, 2014. http://www.creationcare.org.

Evans, John H. "Cooperative Coalitions on the Religious Right and Left: Considering the Resilience of Sectarianism." *Journal for the Scientific Study of Religion* 45, no. 2 (2006): 95–215.

———. "Multi-Organizational Fields and Social Movement Organization Frame Content: The Religious Pro-Choice Movement." *Sociological Inquiry* 67, no. 4 (1997): 451–69.

Evans, Michael S., and John H. Evans. "Arguing against Darwinism: Religion, Science, and Public Morality." In *The New Blackwell Companion to the Sociology of Religion*, edited by Bryan S. Turner, 286–308. Malden, MA: Wiley-Blackwell, 2010.

Farrell, Justin. "The Young and the Restless? The Liberalization of Young Evangelicals." *Journal for the Scientific Study of Religion* 50, no. 3 (2011): 517–32.

Fenn, Richard K. *The Blackwell Companion to the Sociology of Religion.* Malden, MA: Blackwell, 2001.

Finke, Roger. "Innovative Returns to Tradition: Using Core Beliefs as the Foundation for Innovative Accommodation." *Journal for the Scientific Study of Religion* 43 (2004): 19–34.

Fligstein, Neil, and Doug McAdam. "Toward a General Theory of Strategic Action Fields." *Sociological Theory* 29, no. 1 (2011): 1–26.

Fowler, Robert Booth. *The Greening of Protestant Thought.* Chapel Hill: University of North Carolina Press, 1995.

Freudenberg, Nicholas, and Carol Steinsapir. "Not in Our Backyards: The Grassroots Environmental Movement." In *American Environmentalism: The U.S. Environmental Movement, 1970–1990*, edited by Riley E. Dunlap and Angela G. Mertig, 27–37. Philadelphia: Taylor and Francis, 1992.

Friedland, Roger, and Robert R. Alford. "Bringing Society Back In: Symbols, Practices, and Institutional Contradictions." In *The New Institutionalism in Organizational Analysis*, edited by Walter W. Powell and Paul J. DiMaggio. Chicago: University of Chicago Press, 1991.

Fuhse, Jan A. "The Meaning Structure of Social Networks." *Sociological Theory* 27, no. 1 (2009): 51–73.

Fulton, Brad, and Richard L. Wood. "Interfaith Community Organizing: Emerging Theological and Organizational Challenges." *International Journal of Public Theology* 6, no. 4 (2012): 398–420.

Ganz, Marshall. *Why David Sometimes Wins: Leadership, Organization, and Strategy in the California Farm Worker Movement.* New York: Oxford University Press, 2009.

Gardner, Gary. *Invoking the Spirit: Religion and Spirituality in the Quest for a Sustainable World.* Worldwatch Paper No. 164. Washington, DC: Worldwatch Institute, 2002.

Geertz, Clifford. *The Interpretation of Cultures: Selected Essays.* New York: Basic Books, 1973.

Georgia Interfaith Power and Light. "Holiday Resources." Accessed January 30, 2014. http://www.gipl.org/Content/Holiday_Resources.asp.

Gibson, James William. *A Reenchanted World: The Quest for a New Kinship with Nature.* New York: Metropolitan Books, 2009.

Gilham, Patrick F., and Bob Edwards. "Legitimacy, Management, Preservation of Exchange Relationships, and the Dissolution of the Mobilization for Global Justice Campaign." *Social Problems* 58, no. 3 (2011): 433–60.

Goldstone, Jack A. "More Social Movements or Fewer? Beyond Political Opportunity Structure to Relational Fields." *Theory and Society* 33, no. 3–4 (2004): 333–65.

Goodwin, Jeff, and James M. Jasper. "Caught in a Winding, Snarling Vine: The Structural Bias of Political Process Theory." *Sociological Forum* 14, no. 1 (1999): 27–53.

———, eds. *Contention in Context: Political Opportunities and the Emergence of Protest.* Stanford, CA: Stanford University Press, 2012.

Gottlieb, Robert. *Forcing the Spring: The Transformation of the American Environmental Movement*. Washington, DC: Island Press, 1993.

Gottlieb, Roger S. *A Greener Faith: Religious Environmentalism and Our Planet's Future*. New York: Oxford University Press, 2006.

Gould, Roger, and Roberto M. Fernandez. "Structures of Mediation: A Formal Approach to Brokerage in Transaction Networks." *Sociological Methodology* 19 (1989): 89–126.

Granovetter, Mark. "Economic Action and Social Structure: The Problem of Embeddedness." *American Journal of Sociology* 91, no. 3 (1985): 481–510.

Green, Adam Isaiah. "Gay and Lesbian Liberation." In *Contention in Context: Political Opportunities and the Emergence of Protest*, edited by Jeff Goodwin and James M. Jasper. Stanford, CA: Stanford University Press, 2012.

Green, John C. "Evangelical Protestants and Civic Engagement: An Overview." In *A Public Faith: Evangelicals and Civic Engagement*, edited by Michael Cromartie, 11–31. Lanham, MD: Rowman and Littlefield, 2003.

GreenFaith Interfaith Partners for the Environment. "Mission and Areas of Focus." Accessed July 22, 2013. http://www.greenfaith.org/about/mission-and-areas-of-focus.

Greenwood, Royston, and Roy Suddaby. "Institutional Entrepreneurship in Mature Fields: The Big Five Accounting Firms." *Academy of Management Journal* 49, no. 1 (2006): 27–48.

Greenwood, Royston, Roy Suddaby, and C. R. Hinings. "Theorizing Change: The Role of Professional Associations in the Transformation of Institutional Fields." *Academy of Management Journal* 45, no. 1 (2002): 58–80.

Gulati, Ranjay, and Martin Gargiulo. "Where Do Interorganizational Networks Come From?" *American Journal of Sociology* 104 (1999): 1439–93.

Gushee, David P. "Environmental Problem Now a Megathreat." *Creation Care* 26 (Fall 2004): 8–9.

Guth, James L., John C. Green, Lyman A. Kellstedt, and Corwin E. Smidt. "Faith and the Environment: Religious Beliefs and Attitudes on Environmental Policy." *American Journal of Political Science* 39, no. 2 (1995): 364–82.

Guth, James L., Lyman A. Kellstedt, Corwin E. Smidt, and John C. Green. "Theological Perspectives and Environmentalism among Religious Activists." *Journal for the Scientific Study of Religion* 32, no. 4 (1993): 373–82.

Hager, Mark A., Joseph Galaskiewicz, and Jeff A. Larson. "Structural Embeddedness and the Liability of Newness among Nonprofit Organizations." *Public Management Review* 6, no. 2 (2004): 159–88.

Haight, Roger, and John Langan. "Recent Catholic Social and Ethical Teachings in Light of the Social Gospel." *Journal of Religious Ethics* 18, no. 1 (1990): 103–28.

Hall, Charles F. "The Christian Left: Who Are They and How Are They Different from the Christian Right?" *Review of Religious Research* 39, no. 1 (1997): 27–45.

Hannigan, John A. "Social Movement Theory and the Sociology of Religion: Towards a New Synthesis." *Sociological Analysis* 52, no. 4 (1991): 311–31.

Hart, Stephen. *What Does the Lord Require?: How American Christians Think about Economic Justice*. New Brunswick, NJ: Rutgers University Press, 1996.

Hathaway, Will, and David S. Meyer. "Competition and Cooperation in Social Movements." *Berkeley Journal of Sociology* 38 (1994): 157–83.

Haydu, Jeffrey. "Business Citizenship at Work: Cultural Transposition and Class Formation in Cincinnati, 1870–1910." *American Journal of Sociology* 107, no. 6 (2002): 1424–67.

Hazon. "Overview—2022 Vision: 7-Year Goals for the Jewish Food Movement." Accessed January 15, 2014. http://hazon.org/jewish-food-movement/overview.

Heaney, Michael T., and Fabio Rojas. "Hybrid Activism: Social Movement Mobilization in a Multi-Movement Environment." *American Journal of Sociology* 119, no. 4 (2014): 1047–103.

Hessel, Dieter T., and Rosemary Radford Ruether, eds. *Christianity and Ecology: Seeking the Well-Being of Earth and Humans.* Cambridge, MA: Harvard University Press, 2000.

Hojnacki, Marie. "Interest Groups' Decisions to Join Alliances or Work Alone." *American Journal of Political Science* 41 (1997): 61–87.

Hsu, Greta, and Michael T. Hannan. "Identities, Genres, and Organizational Forms." *Organization Science* 16, no. 5 (2005): 474–90.

Jasper, James M. "The Innovation Dilemma: Some Risks of Creativity in Strategic Action." In *The Dark Side of Creativity*, edited by David H. Cropley, Arthur J. Cropley, James C. Kaufman, and Mark A. Runco, 91–113. New York: Cambridge University Press, 2010.

———. "Introduction: From Political Opportunity Structures to Strategic Action." In *Contention in Context: Political Opportunities and the Emergence of Protest*, edited by Jeff Goodwin and James M. Jasper. Stanford, CA: Stanford University Press, 2012.

———. "A Strategic Approach to Collective Action: Looking for Agency in Social Movement Choices." *Mobilization* 9, no. 1 (2004): 1–16.

Johnston, Hank, Enrique Larana, and Joseph R. Gusfield. "Identities, Grievances, and New Social Movements." In *New Social Movements: From Ideology to Identity*, edited by Enrique Larana, Hank Johnston, and Joseph R. Gusfield, 3–35. Philadelphia: Temple University Press, 1994.

Kanagy, Conrad L., and Hart M. Nelson. "Religion and Environmental Concern: Challenging the Dominant Assumptions." *Review of Religious Research* 37, no. 1 (1995): 33–45.

Kearns, Laurel. "Ecology and the Environment." In *The Blackwell Companion to Religion and Social Justice*, edited by Michael D. Palmer and Stanley M. Burgess, 591–606. Malden, MA: Wiley-Blackwell, 2012.

———. "Noah's Ark Goes to Washington: A Profile of Evangelical Environmentalism." *Social Compass* 44, no. 3 (1997): 349–66.

———. "Saving the Creation: Christian Environmentalism in the United States." *Sociology of Religion* 57, no. 1 (1996): 55–70.

Kellestedt, Lyman A., and John C. Green. "The Politics of the Willow Creek Association." *Journal for the Scientific Study of Religion* 42, no. 4 (2003): 547–61.

Kern, Thomas. "Cultural Performance and Political Regime Change." *Theory and Society* 27, no. 3 (2009): 291–316.

Kitschelt, Herbert. "Political Opportunity Structures and Political Protest: Anti-Nuclear Movements in Four Democracies." *British Journal of Political Science* 16, no. 1 (1986): 57–85.

Klandermans, Bert, Jojanneke van der Toorn, and Jacquelin van Stekelenburg. "Embeddedness and Identity: How Immigrants Turn Grievances into Action." *American Sociological Review* 73, no. 6 (2008): 992–1012.

Kniss, Fred, and Gene Burns. "Religious Movements." In *The Blackwell Companion to Social Movements*, edited by David A. Snow, Sarah Soule, and Hanspeter Kriesi, 694–715. Malden, MA: Blackwell, 2004.

Kniss, Fred, and David Todd Campbell. "The Effect of Religious Orientation on International Relief and Development Organizations." *Journal for the Scientific Study of Religion* 36, no. 1 (1997): 93–103.

Krauss, Rachel. "Laity, Institution, Theology, or Politics? Protestant, Catholic, and Jewish Washington Offices' Agenda Setting." *Sociology of Religion* 68, no. 1 (2007): 67–81.

Kucinskas, Jaime. "The Unobtrusive Tactics of Religious Movements." *Sociology of Religion* 75, no. 4 (2014): 537–50.

Lehman, David. "Charisma and Possession in Africa and Brazil." *Theory, Culture, and Society* 18, no. 5 (2001): 45–74.

Levi, Margaret, and Gillian H. Murphy. "Coalitions of Contention: The Case of WTO Protests in Seattle." *Political Studies* 54, no. 4 (2006): 651–70.

Levitsky, Sandra R. "Niche Activism: Constructing a Unified Movement Identity in a Heterogeneous Organizational Field." *Mobilization* 12, no. 3 (2007): 271–86.

Lichterman, Paul. *Elusive Togetherness: Church Groups Trying to Bridge America's Divisions.* Princeton, NJ: Princeton University Press, 2005.

———. "Piecing Together Multicultural Community: Cultural Differences in Community Building among Grass-Roots Environmentalists." *Social Problems* 42, no. 4 (1995): 513–34.

———. *The Search for Political Community: American Activists Reinventing Commitment.* New York: Cambridge University Press, 1996.

Lindsay, D. Michael. "Evangelicals in the Power Elite: Elite Cohesion Advancing a Movement." *American Sociological Review* 73, no. 1 (2008): 60–82.

Lounsbury, Michael, Marc Ventresca, and Paul M. Hirsch. "Social Movements, Field Frames, and Industry Emergence: A Cultural-Political Perspective on US Recycling." *Socio-Economic Review* 1, no. 1 (2003): 71–104.

Maguire, Steve, Cynthia Hardy, and Thomas B. Lawrence. "Institutional Entrepreneurship in Emerging Fields: HIV/AIDS Treatment Advocacy in Canada." *Academy of Management Journal* 47, no. 5 (2004): 657–79.

Marx, Karl. *The Eighteenth Brumaire of Louis Bonaparte.* New York: International Publishers, 1987 [1869].

McAdam, Doug. *Political Process and the Development of Black Insurgency, 1930–1970.* Chicago: University of Chicago Press, 1982.

McAdam, Doug, John D. McCarthy, and Mayer N. Zald. "Introduction: Opportunities, Mobilizing Structures, and Framing Processes—Toward a Synthetic, Comparative Perspective on Social Movements." In *Comparative Perspectives on Social Movements: Political Opportunities, Mobilizing Structures, and Cultural Framings,* edited by Doug McAdam, John D. McCarthy, and Mayer N. Zald, 1–20. Cambridge: Cambridge University Press, 1996.

———, eds. *Comparative Perspectives on Social Movements: Political Opportunities, Mobilizing Structures, and Cultural Framings.* Cambridge: Cambridge University Press, 1996.

McCammon, Holly, and Karen E. Campbell. "Allies on the Road to Victory: Coalition Formation between the Suffragists and the Women's Christian Temperance Union." *Mobilization* 7 (2002): 231–51.

McCammon, Holly, Soma Chaudhuri, Lyndi Hewitt, Courtney Sanders Muse, Harmony D. Newman, Carrie Lee Smith, and Teresa M. Terrell. "Becoming Full Citizens: The U.S. Women's Jury Rights Campaigns, the Pace of Reform, and Strategic Adaptation." *American Journal of Sociology* 113, no. 4 (2008): 1104–47.

McCammon, Holly J., and Nella Van Dyke. "Applying Qualitative Comparative Analysis to Empirical Studies of Social Movement Coalition Formation." In *Strategic Alliances: Coalition Building and Social Movements,* edited by Nella Van Dyke and Holly J. McCammon, 292–315. Minneapolis: University of Minnesota Press, 2010.

Meyer, David S. "Protest and Political Opportunities." *Annual Review of Sociology* 30 (2004): 125–45.

Meyer, David S., and Catherine Corrigall-Brown. "Coalitions and Political Context: U.S. Movements against Wars in Iraq." *Mobilization* 10, no. 3 (2005): 327–44.

Meyer, David S., and Debra C. Minkoff. "Conceptualizing Political Opportunity." *Social Forces* 82, no. 4 (2004): 1457–92.

Meyer, J. W., and W. R. Scott. "Centralization and the Legitimacy Problems of the Local Government." In *Organizational Environments: Ritual and Rationality*, edited by John W. Meyer and W. Richard Scott, 199–215. Beverly Hills, CA: Sage, 1983.

Michigan Interfaith Power and Light. Accessed August 22, 2007. http://www.miipl.org.

Miller, Danny. "The Embeddedness of Corporate Strategy: Isomorphism vs. Differentiation." *Advances in Strategic Management* 13 (1996): 283–91.

Miller, Donald E. *Reinventing American Protestantism: Christianity in the New Millennium.* Berkeley: University of California Press, 1997.

Mirola, William A. "Religious Protest and Economic Conflict: Possibilities and Constraints on Religious Resources Mobilization and Coalitions in Detroit's Newspaper Strike." *Sociology of Religion* 64, no. 4 (2003): 443–61.

Moore, Kelly. *Disrupting Science: Social Movements, American Scientists, and the Politics of the Military, 1945–1975.* Princeton, NJ: Princeton University Press, 2008.

Morphew, Christopher C., and Matthew Hartley. "Mission Statements: A Thematic Analysis of Rhetoric across Institutional Type." *Journal of Higher Education* 77, no. 3 (2006): 456–71.

Morris, Aldon D. *The Origins of the Civil Rights Movement: Black Communities Organizing for Change.* New York: Free Press, 1984.

Munson, Ziad W. *The Making of Pro-Life Activists: How Social Movement Mobilization Works.* Chicago: University of Chicago Press, 2008.

Murray, Fiona. "The Oncomouse That Roared: Hybrid Exchange Strategies as a Source of Distinction at the Boundary of Overlapping Institutions." *American Journal of Sociology* 116, no. 2 (2010): 341–88.

Myerhoff, Barbara. *Number Our Days: A Triumph of Continuity and Culture among Jewish Old People in an Urban Ghetto.* New York: Simon and Schuster, 1978.

Nepstad, Sharon Erickson. *Convictions of the Soul: Religion, Culture, and Agency in the Central American Solidarity Movement.* New York: Oxford University Press, 2004.

———. "Narrative in the U.S./Central American Peace Movement." *Mobilization* 6, no. 1 (2001): 21–36.

———. "Oppositional Consciousness among the Privileged: Remaking Religion in the Central America Solidarity Movement." *Critical Sociology* 33, no. 4 (2007): 661–88.

Nichols, Joel A. "Evangelicals and Human Rights: The Continuing Ambivalence of Evangelical Christians' Support for Human Rights." *Journal of Law and Religion* 24, no. 2 (2008–9): 629–62.

Offe, Claus. "New Social Movements: Challenging the Boundaries of Institutional Politics." *Social Research* 52, no. 4 (1985): 817–68.

Orbach, Brian K. *Labor and the Environmental Movement: A Quest for Common Ground.* Cambridge, MA: MIT Press, 2004.

Pattillo-McCoy, Mary. "Church Culture as a Strategy of Action in the Black Community." *American Sociological Review* 63, no. 6 (1998): 767–84.

Polletta, Francesca. "Culture and Movements." *ANNALS of the American Academy of Political and Social Science* 619 (September 2008): 78–96.

Polody, Joel, and Karen L. Page. "Network Forms of Organization." *Annual Review of Sociology* 24 (1998): 57–76.

Potter, Samuel C. "The Pacific Northwest Forest Debate: Bringing Religion Back In?" *World-views* 3, no. 1 (1999): 3–32.

Powell, Walter W. "Neither Market nor Hierarchy: Network Forms of Organization." *Research in Organizational Behavior* 12 (1990): 295–336.

Rao, Hayagreeva. "Caveat Emptor: The Construction of Nonprofit Consumer Watchdog Organizations." *American Journal of Sociology* 103, no. 4 (1998): 912–61.

Rao, Hayagreeva, and Simona Giorgi. "Code Breaking: How Entrepreneurs Exploit Cultural Logics to Generate Institutional Change." *Research in Organizational Behavior* 27 (2006): 269–304.

Rao, Hayagreeva, Philippe Monin, and Rodolphe Durand. "Border Crossing: Bricolage and the Erosion of Categorical Boundaries in French Gastronomy." *American Sociological Review* 70, no. 6 (2005): 968–91.

Rao, Hayagreeva, Calvin Morrill, and Mayer N. Zald. "Power Plays: How Collective Movements and Collective Action Create New Organizational Forms." *Research in Organizational Behavior* 22 (2000): 237–81.

Restoring Eden. "History: Restoring Eden." Accessed May 15, 2013. http://restoringeden.org/about/history.

———. "An open letter to Dobson concerning climate change." Accessed February 25, 2014. http://restoringeden.org/connect/CreationVoice/2007/OpenLetter.

Rochon, Thomas R., and David S. Meyer, eds. *Coalitions and Political Movements: The Lessons of the Nuclear Freeze.* Boulder, CO: Lynne Reiner, 1997.

Rohlinger, Deanna, and Jill Quadagno. "Framing Faith: Cooperation and Conflict in the U.S. Conservative Christian Political Movement." *Social Movement Studies* 8, no. 4 (2009).

Roof, Wade Clark. *Spiritual Marketplace: Baby Boomers and the Remaking of American Religion.* Princeton, NJ: Princeton University Press, 1999.

Rose, Fred. *Coalitions across the Class Divide: Lessons from the Labor, Peace, and Environmental Movements.* Ithaca, NY: Cornell University Press, 2000.

Rosenthal, Gilbert S. "*Tikkun ha-Olam*: The Metamorphosis of a Concept." *Journal of Religion* 85, no. 2 (2005): 214–40.

Roth, Benita. "'Organizing on One's Own' as Good Politics: Second Wave Feminists and the Meaning of Coalition." In *Strategic Alliances: Coalition Building and Social Movements*, edited by Nella Van Dyke and Holly J. McCammon, 99–118. Minneapolis: University of Minnesota Press, 2010.

Ruef, Martin. "The Emergence of Organizational Forms: A Community Ecology Approach." *American Journal of Sociology* 106, no. 3 (2000): 658–714.

Savage, Nigel. "Keeping Kosher in the Time of McDonalds and Monsanto: American Jews and the Sacred Food Movement." *National Catholic Rural Life Conference Magazine* (Winter 2006).

Sewell, William H., Jr. "A Theory of Structure: Duality, Agency, and Transformation." *American Journal of Sociology* 98, no. 1 (1992): 1–29.

Shalom Center, The. Accessed January 30, 2014. https://theshalomcenter.org.

Shellenberger, Michael, and Ted Nordhaus. "The Death of Environmentalism: Global Warming in a Post-Environmental World." Accessed May 22, 2015. http://www.thebreakthrough.org/images/Death_of_Environmentalism.pdf.

Sherkat, Darren E., and Christopher G. Ellison. "Structuring the Religion-Environmental Connection: Identifying Religious Influences on Environmental Concern and Activism." *Journal for the Scientific Study of Religion* 46, no. 1 (2007): 71–85.

Shibley, Mark A., and Jonathan L. Wiggins. "The Greening of Mainline American Religion:

A Sociological Analysis of the Environmental Ethics of the National Religious Partnership for the Environment." *Social Compass* 44, no. 3 (1997): 333–48.

Smith, Angela M., and Simone Pulver. "Ethics-Based Environmentalism in Practice: Religious-Environmental Organizations in the United States." *Worldviews* 13, no. 2 (2009): 145–79.

Smith, Buster G., and Bryon Johnson. "The Liberalization of Young Evangelicals: A Research Note." *Journal for the Scientific Study of Religion* 49, no. 2 (2010): 351–60.

Smith, Christian. *American Evangelicalism: Embattled and Thriving.* Chicago: University of Chicago Press, 1998.

———. *Disruptive Religion: The Force of Faith in Social-Movement Activism.* New York: Routledge, 1996.

———. *Resisting Reagan: The U.S. Central America Peace Movement.* Chicago: University of Chicago Press, 1996.

Smith, Steven Rathgeb, and Michael R. Sosin. "The Varieties of Faith-Related Agencies." *Public Administrative Review* 61, no. 6 (2001): 651–70.

Snow David A., E. Burke Rochford, Steven K. Worden, and Robert D. Benford. "Frame Alignment Processes, Micromobilization, and Movement Participation." *American Sociological Review* 51, no. 4 (1986): 464–81.

Snow, David A., and Sarah A. Soule. *A Primer on Social Movements.* New York: W. W. Norton, 2010.

Snow, David A., Sarah A. Soule, and Hanspeter Kriesi, eds. *The Blackwell Companion to Social Movements.* Malden, MA: Blackwell, 2004.

Snow, David, Anne E. Tan, and Peter B. Owens. "Social Movements, Framing Processes, and Cultural Revitalization and Fabrication." *Mobilization* 18, no. 3 (2013): 225–42.

Somplatsky-Jarman, William, Walt E. Grazer, and Stan L. LeQuire. "Partnership for the Environment among U.S. Christians: Reports from the National Religious Partnership for the Environment." In *Christianity and Ecology: Seeking the Well-Being of Earth and Humans,* edited by Dieter T. Hessel and Rosemary Radford Ruether, 573–90. Cambridge, MA: Harvard University Press, 2000.

Southern Baptist Environment and Climate Initiative. "A Southern Baptist Declaration on the Environment and Climate Change." Accessed April 9, 2015. http://www.baptistcreationcare.org/node/1.

Stadler, Nurit. "Is Profane Work an Obstacle to Salvation? The Case of Ultra Orthodox (Haredi) Jews in Contemporary Israel." *Sociology of Religion* 63, no. 4 (2002): 455–74.

Staggenborg, Suzanne. "Coalition Work in the Pro-Choice Movement: Organizational and Environmental Opportunities and Constraints." *Social Problems* 33 (1986): 374–80.

———. "Conclusion: Research on Social Movement Coalitions." In *Strategic Alliances: Coalition Building and Social Movements,* edited by Nella Van Dyke and Holly J. McCammon, 316–29. Minneapolis: University of Minnesota Press, 2010.

Steensland, Brian, and Phillip Goff. *The New Evangelical Engagement.* New York: Oxford University Press, 2014.

Strang, David, and John W. Meyer. "Institutional Conditions for Diffusion." *Theory and Society* 22, no. 4 (1993): 487–511.

Suchman, M. C. "Managing Legitimacy: Strategic and Institutional Approaches." *Academy of Management Review* 20 (1995): 571–611.

Suddaby, Roy, and Royston Greenwood. "Rhetorical Strategies of Legitimacy." *Administrative Science Quarterly* 50, no. 1 (2005): 35–67.

Swidler, Ann. "Culture in Action: Symbols and Strategies." *American Sociological Review* 51, no. 2 (1986): 273–86.

Taylor, Bron. "Earth and Nature-Based Spirituality (Part I): From Deep Ecology to Radical Environmentalism." *Religion* 31, no. 2 (2001): 175–93.

———. "Earth and Nature-Based Spirituality (Part II): From Earth First! and Bioregionalism to Scientific Paganism and the New Age." *Religion* 31, no. 3 (2001): 225–45.

———. "Wilderness, Spirituality, and Biodiversity in North America—Tracing an Environmental History from Occidental Roots to Earth Day." In *Wilderness in Mythology and Religion: Approaching Religious Spatialities, Cosmologies, and Ideas of Wild Nature*, edited by Laura Feldt, 293–324. New York: De Gruyter, 2012.

Taylor, Sarah McFarland. *Green Sisters: A Spiritual Ecology*. Cambridge, MA: Harvard University Press, 2007.

Tucker, Mary Evelyn. "Religion and Ecology." In *The Oxford Handbook of the Sociology of Religion*, edited by Peter B. Clarke, 819–35. New York: Oxford University Press, 2009.

Unitarian Universalist Association. "What We Believe." Accessed January 17, 2014. http://www.uua.org/beliefs/what-we-believe.

Van Dyke, Nella. "Crossing Movement Boundaries: Factors That Facilitate Coalition Protest by American College Students, 1930–1990." *Social Problems* 50 (2003): 226–50.

Van Dyke, Nella, and Holly J. McCammon. "Introduction: Social Movement Coalition Formation." In *Strategic Alliances: Coalition Building and Social Movements*, edited by Nella Van Dyke and Holly J. McCammon, xi–xxviii. Minneapolis: University of Minnesota Press, 2010.

———, eds. *Strategic Alliances: Coalition Building and Social Movements*. Minneapolis: University of Minnesota Press, 2010.

Van Geest, Fred. "Changing Patterns of Denominational Activity in North America: The Case of Homosexuality." *Review of Religious Research* 49, no. 2 (2007): 199–221.

Vatican, The. "Message of His Holiness Pope John Paul II for the Celebration of the World Day of Peace." Accessed March 10, 2014. http://w2.vatican.va/content/john-paul-ii/en/messages/peace/documents/hf_jp-ii_mes_19891208_xxiii-world-day-for-peace.html.

Warner, Keith Douglass, O.F.M. "The Greening of American Catholicism: Identity, Conversion, and Continuity." *Religion and American Culture* 18, no. 1 (2008): 113–42.

Warner, R. Stephen. "Work in Progress toward a New Paradigm for the Sociological Study of Religion in the United States." *American Journal of Sociology* 98, no. 5 (1993): 1044–93.

Warren, Mark R. *Dry Bones Rattling: Community Building to Revitalize American Democracy*. Princeton, NJ: Princeton University Press, 2001.

Weiss, Janet A., and Sandy Dritin Piderit. "The Value of Mission Statements in Public Agencies." *Journal of Public Administration and Theory: J-Part* 9, no. 2 (1999): 193–223.

Wellman, James K. *Evangelical vs. Liberal: The Clash of Christian Cultures in the Pacific Northwest*. New York: Oxford University Press, 2008.

Wilford, Justin G. *Sacred Subdivisions: The Postsuburban Transformation of American Evangelicalism*. New York: New York University Press, 2012.

Wilkinson, Katharine K. *Between God and Green: How Evangelicals Are Cultivating a Middle Ground on Climate Change*. New York: Oxford University Press, 2012.

Williams, Johnny E. "Linking Beliefs to Collective Action: Politicized Religious Beliefs and the Civil Rights Movement." *Sociological Forum* 17, no. 2 (2002): 203–22.

Williams, Rhys. "Constructing the Public Good: Social Movements and Cultural Resources." *Social Problems* 42, no. 1 (1995): 124–44.

———. "From the 'Beloved Community' to 'Family Values': Religious Language, Symbolic Repertoires, and Democratic Culture." In *Social Movements: Identity, Culture, and the*

State, edited by David S. Meyer, Nancy Whittier, and Belinda Robnett, 247–65. New York: Oxford University Press, 2002.

———. "Religious Social Movements in the Public Sphere: Organization, Ideology, and Activism." In *Handbook of the Sociology of Religion,* edited by Michele Dillon, 315–30. Cambridge: Cambridge University Press, 2003.

Williams, Rhys H., and Jeffrey Blackburn. "Many Are Called but Few Obey: Ideological Commitment and Activism in Operation Rescue." In *Disruptive Religion: The Force of Faith in Social-Movement Activism,* edited by Christian Smith, 167–85. New York: Routledge, 1996.

Wolkomir, Michelle, Eric Woodrum, Michael Futreal, and Thomas Hoban. "Denominational Subcultures of Environmentalism." *Review of Religious Research* 38, no. 4 (1997): 325–43.

Wood, Richard L. *Faith in Action: Religion, Race, and Democratic Organizing in America.* Chicago: University of Chicago Press, 2002.

Wuthnow, Robert. *After the Baby Boomers: How Twenty- and Thirty-Somethings Are Shaping the Future of American Religion.* Princeton, NJ: Princeton University Press, 2007.

———. *Meaning and Moral Order: Explorations in Cultural Analysis.* Berkeley: University of California Press, 1987.

———. *The Restructuring of American Religion: Society and Faith since World War II.* Princeton, NJ: Princeton University Press, 1988.

———. *Saving America: Faith-Based Services and the Future of Civil Society.* Princeton, NJ: Princeton University Press, 2004.

Yamane, David. *The Catholic Church in State Politics: Negotiating Prophetic Demands and Political Realities.* Lanham, MD: Rowman and Littlefield, 2005.

Young, Michael P. *Bearing Witness against Sin: The Evangelical Birth of the American Social Movement.* Chicago: University of Chicago Press, 2006.

Zald, Meyer. "Theological Crucibles: Social Movements in and of Religion." *Review of Religious Research* 23, no. 1 (1982): 317–36.

Zuckerman, Ezra. "The Categorical Imperative: Securities Analysts and the Illegitimacy Discount." *American Journal of Sociology* 104, no. 5 (1999): 1398–438.

Zukin, Sharon, and Paul DiMaggio. "Introduction." In *Structures of Capital: The Social Organization of the Economy,* edited by Sharon Zukin and Paul DiMaggio, 1–36. New York: Cambridge University Press, 1990.

INDEX